Proofs Without Words II

More Exercises in Visual Thinking

Proofs Without Words II

More Exercises in Visual Thinking

Roger B. Nelsen
Lewis & Clark College

Published and Distributed by

THE MATHEMATICAL ASSOCIATION OF AMERICA

CLASSROOM RESOURCE MATERIALS

This series provides supplementary material for students and their teachers—laboratory exercises, projects, historical information, textbooks with unusual approaches for presenting mathematical ideas, career information, and much more.

101 Careers in Mathematics, edited by Andrew Sterrett

Archimedes: What Did He Do Besides Cry Eureka?, Sherman Stein

Calculus Mysteries and Thrillers, R. Grant Woods

Combinatorics: A Problem Oriented Approach, Daniel A. Marcus

A Course in Mathematical Modeling, Douglas Mooney and Randall Swift

Cryptological Mathematics, Robert Edward Lewand

Elementary Mathematical Models, Dan Kalman

Geometry From Africa: Mathematical and Educational Explorations, Paulus Gerdes

Interdisciplinary Lively Application Projects, edited by Chris Arney

Laboratory Experiences in Group Theory, Ellen Maycock Parker

Learn from the Masters, Frank Swetz, John Fauvel, Otto Bekken, Bengt Johansson, and Victor Katz

Mathematical Modeling in the Environment, Charles Hadlock

A Primer of Abstract Mathematics, Robert B. Ash

Proofs Without Words, Roger B. Nelsen

Proofs Without Words II, Roger B. Nelsen

A Radical Approach to Real Analysis, David M. Bressoud

She Does Math!, edited by Marla Parker

MAA Service Center
P.O. Box 91112
Washington, DC 20090-1112
1-800-331-1MAA   FAX: 1-301-206-9789

This book is dedicated to the memory of my parents,

Ann Bain Nelsen and Howard Ernest Nelsen.

Contents

Introduction

Proofs really aren't there to convince you that
something is true—they're there to show you why
it is true.

—Andrew Gleason

A good proof is one that makes us wiser.

—Yu. I. Manin

Much research for new proofs of theorems already
correctly established is undertaken simply because
the existing proofs have no aesthetic appeal. There
are mathematical demonstrations that are merely
convincing; to use a phrase of the famous mathematical physicist, Lord Rayleigh, they "command
assent." There are other proofs "which woo and
charm the intellect. They evoke delight and an
overpowering desire to say, Amen, Amen." An ele-
gantly executed proof is a poem in all but the form
in which it is written.

—Morris Kline

What are "proofs without words?" As you will see from this sec-
ond collection, the question does not have a simple, concise answer
(the first collection, *Proofs Without Words: Exercises in Visual Think-
ing*, was published by the Mathematical Association of America in
1993). Generally, proofs without words (PWWs) are pictures or dia-
grams that help the reader see *why* a particular mathematical state-
ment may be true, and also to see *how* one might begin to go about
proving it true. In some, an equation or two may appear in order to
guide the observer in this process. The emphasis is, however, clearly
on providing visual clues to the observer to stimulate mathematical
thought.

Proofs without words are regular features in the journals published
by the Mathematical Association of America. PWWs began to appear in
Mathematics Magazine about 1975, and in *The College Mathematics
Journal* about ten year later. But proofs without words are not recent
innovations—they have been around for a very long time. In this vol-

ume you find modern renditions of PWWs from ancient China, tenth century Arabia, and Renaissance Italy. Proofs without words also now appear in journals published by other organizations in the U.S. and abroad, and on the World Wide Web.

Of course, some argue that PWWs are not really "proofs" (nor, for that matter, are they "without words," on account of equations which often accompany a PWW). In his recent book *Philosophy of Mathematics: An Introduction to the World of Proofs and Pictures* (Routledge, London, 1999), James Robert Brown notes:

> "Mathematicians, like the rest of us, cherish clever ideas; in particular they delight in an ingenious picture. But this appreciation does not overwhelm a prevailing skepticism. After all, a diagram is—at best—just a special case and so can't establish a general theorem. Even worse, it can be downright misleading. Though not universal, the prevailing attitude is that pictures are really no more than heuristic devices; they are psychologically suggestive and pedagogically important—but they *prove* nothing. I want to oppose this view and to make a case for pictures having a legitimate role to play as evidence and justification—a role well beyond the heuristic. In short, pictures can prove theorems."

In my introduction to the first collection of PWWs, I suggested that teachers might want to share the PWWs with their students. Several readers of the first collection responded to my request for information about ways in which PWWs are being used in the classroom. Respondents commented on using PWWs with classes at all levels— precalculus courses in high school, college courses in calculus, number theory, and combinatorics, and pre-service and in-service classes for teachers. PWWs appear to be used frequently to supplement or even replace "textbook" proofs, for example, for the Pythagorean theorem or the formulas for sums of integers, squares, and cubes. Other uses range from regular assignments, extra-credit problems, in-class presentations by students, and even term papers and projects.

I should note that this collection, like the first, is necessarily incomplete. It does not include all PWWs which have appeared in print since the first collection was published in 1993, nor all of those which I overlooked in compiling the first book. As readers of the Association's journals are well aware, new PWWs appear in print rather frequently, and they also appear now on the World Wide Web in formats superior to print, involving motion and viewer interaction.

I hope that the readers of this collection will find enjoyment in discovering or rediscovering some elegant visual demonstrations of certain mathematical ideas; that teachers will share them with their students; and that all will find stimulation and encouragement to create new proofs without words.

Acknowledgment. I would like to express my appreciation and gratitude to all those individuals who have contributed proofs without words to the mathematical literature; see the *Index of Names* on pp. 127-128. Without them this collection simply would not exist. Thanks to Andy Sterrett and the members of the editorial board of Classroom Resource Materials for their careful reading of an earlier draft of the book, and for their many helpful suggestions. I would also like to thank Elaine Pedreira, Beverly Ruedi, and Don Albers of the MAA's publication staff for their encouragement, expertise, and hard work in preparing this book for publication.

Roger B. Nelsen
Lewis & Clark College
Portland, Oregon

Notes

1. The illustrations in this collection were redrawn to create a uniform appearance. In a few instances titles were changed, and shading or symbols were added or deleted for clarity. Any errors resulting from that process are entirely my responsibility.

2. Roman numerals are used in the titles of some PWWs to distinguish multiple PWWs of the same theorem—and the numbering is carried over from *Proofs Without Words*. So, for example, since there are six PWWs of the Pythagorean Theorem in *Proofs Without Words*, the first in this collection carries the title "The Pythagorean Theorem VII."

3. Several PWWs in this collection are presented in the form of "solutions" to problems from mathematics contests such as the William Lowell Putnam Mathematical Competition and the Canadian Mathematical Olympiad. It is quite doubtful that such "solutions" would have garnered many points in those contests, as contestants are advised in, for example, the Putnam Competition that "all the necessary steps of a proof must be shown clearly to obtain full credit."

4. The three quotations at the beginning of the Introduction are from *Out of the Mouths of Mathematicians* by Rosemary Schmalz, (Mathematical Association of America, Washington, 1993), pp. 75, 62, and 135-136.

Geometry & Algebra

The Pythagorean Theorem VII

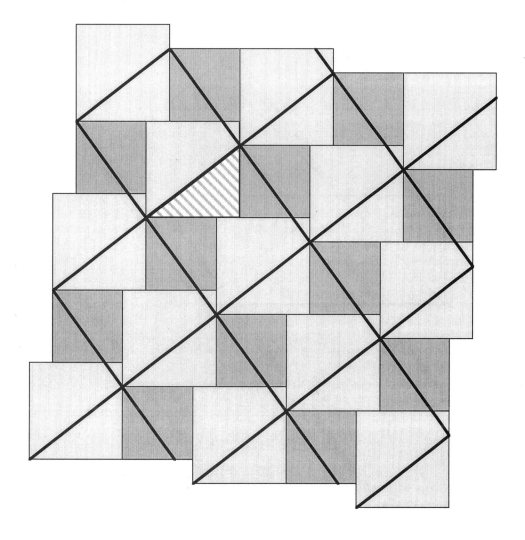

—Annairizi of Arabia (circa A.D. 900)

The Pythagorean Theorem VIII

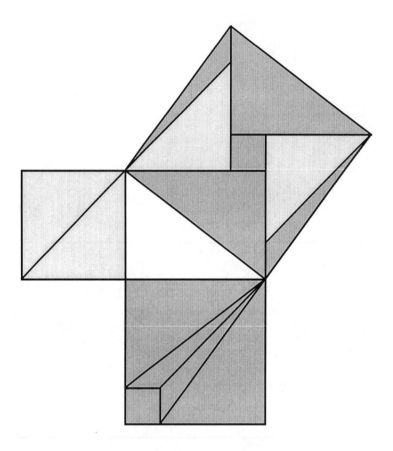

—Liu Hui (3rd century A.D.)

The Pythagorean Theorem IX

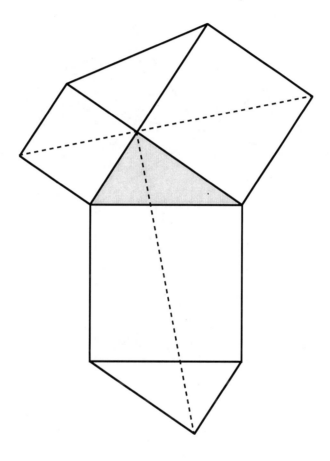

—Leonardo da Vinci (1452-1519)

The Pythagorean Theorem X

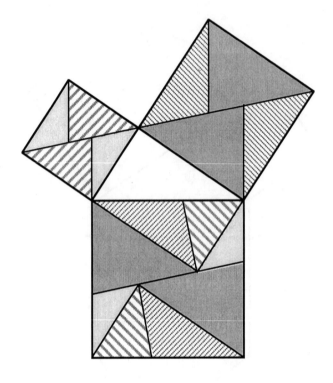

—J. E. Böttcher

The Pythagorean Theorem XI

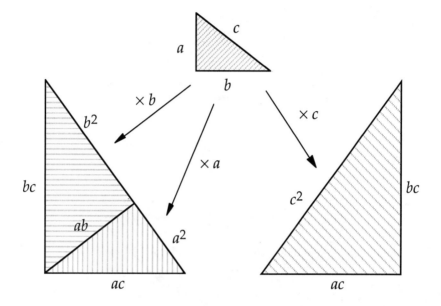

—Frank Burk

The Pythagorean Theorem XII

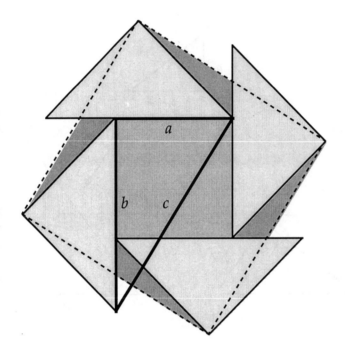

$$a^2 + b^2 = c^2$$

—Poo-sung Park

A Generalization from Pythagoras

The sum of the area of two squares, whose sides are the lengths of the two diagonals of a parallelogram, is equal to the sum of the areas of four squares, whose sides are its four sides.

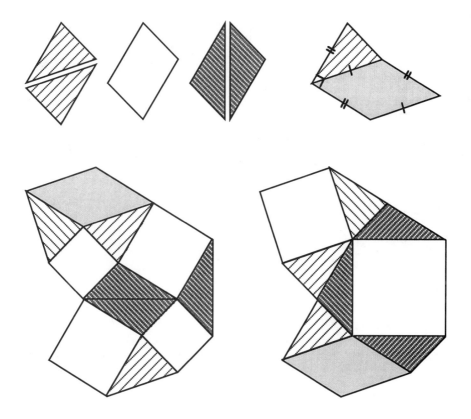

COROLLARY: The Pythagorean Theorem (when the parallelogram is a rectangle).

—David S. Wise

A Theorem of Hippocrates of Chios (circa B.C. 440)

The combined area of the lunes constructed on the legs of a given right triangle is equal to the area of the triangle.

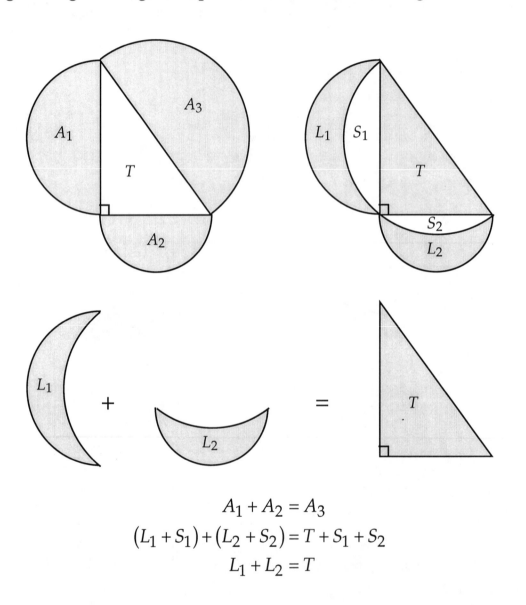

$$A_1 + A_2 = A_3$$
$$(L_1 + S_1) + (L_2 + S_2) = T + S_1 + S_2$$
$$L_1 + L_2 = T$$

—Eugene A. Margerum
and Michael M. McDonnell

The Area of a Right Triangle with Acute Angle $\pi/12$

The area of a right triangle is $\frac{1}{8}$(hypotenuse)^2 if and only if one acute angle is $\pi/12$.

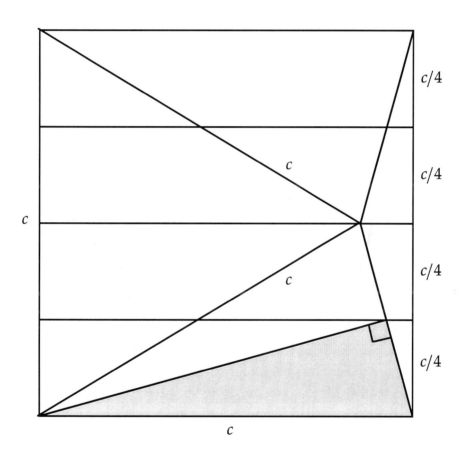

—Klara Pinter

A Right Triangle Inequality

(Problem 3, The Canadian Mathematical Olympiad, 1969)

Let c be the length of the hypotenuse of a right triangle whose other two sides have lengths a and b. Prove that

$$a + b \leq c\sqrt{2}.$$

When does equality hold?

SOLUTION:

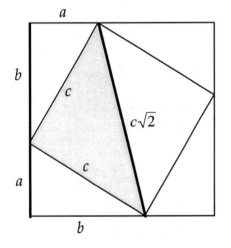

$$a + b \leq c\sqrt{2}$$

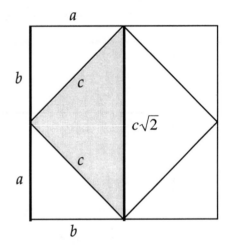

$$a + b = c\sqrt{2} \Leftrightarrow a = b$$

The Inradius of a Right Triangle

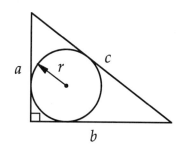

I. $r = \dfrac{ab}{a+b+c}$

II. $r = \dfrac{a+b-c}{2}$

I. $ab = r(a+b+c)$

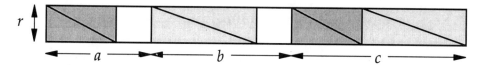

II. $c = a + b - 2r$

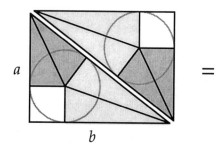

—Liu Hui (3rd century A.D.)

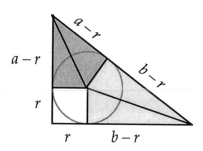

The Product of the Perimeter of a Triangle and Its Inradius is Twice the Area of the Triangle

I.

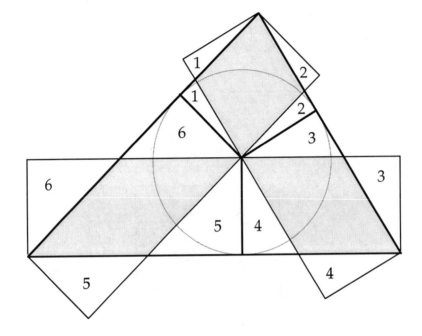

NOTE: Regions bearing the same number are equal in area.

—Grace Lin

II.

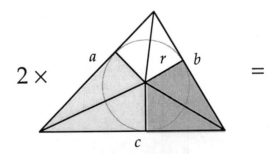

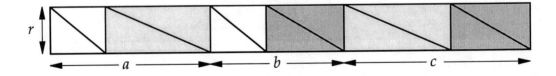

Four Triangles with Equal Area

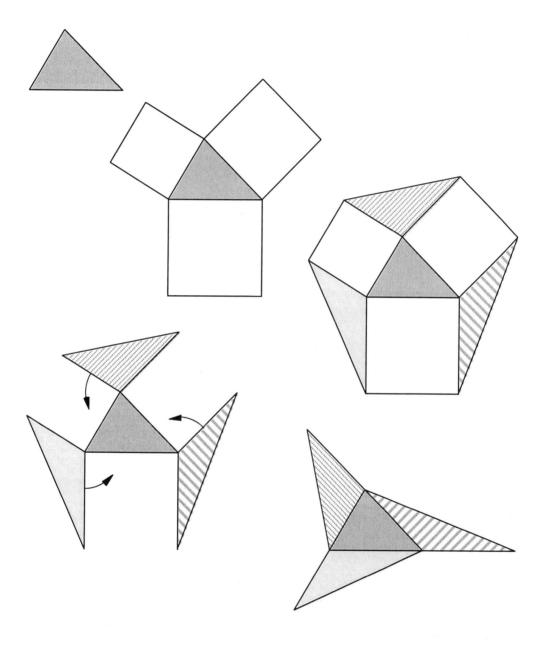

—Steven L. Snover

The Triangle of Medians Has Three-Fourths the Area of the Original Triangle

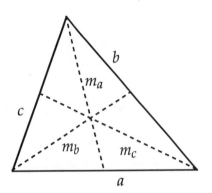 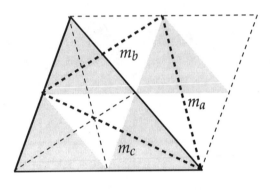

$$\text{area}\left(\Delta m_a m_b m_c\right) = \frac{3}{4}\,\text{area}\left(\Delta abc\right)$$

—Norbert Hungerbühler

Heptasection of a Triangle

If the one-third points on each side of a triangle are joined to opposite vertices, the resulting central triangle is equal in area to one-seventh that of the initial triangle.

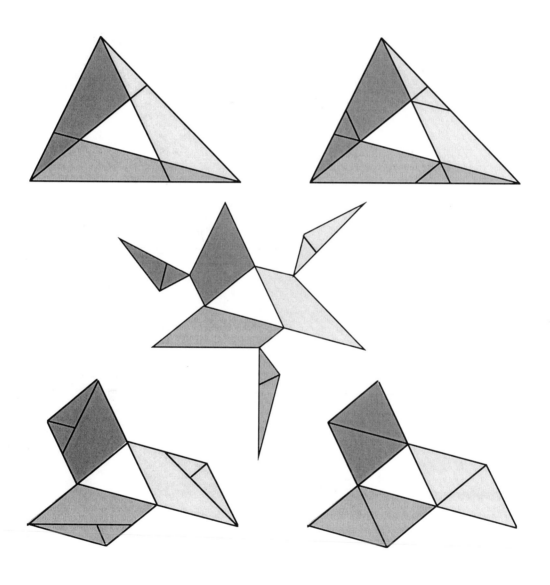

—William Johnston
and Joe Kennedy

A Golden Section Problem from the *Monthly*

(Problem E3007, *American Mathematical Monthly*, 1983, p. 482)

Let *A* and *B* be the midpoints of the sides *EF* and *ED* of an equilateral triangle *DEF*. Extend *AB* to meet the circumcircle (of *DEF*) at *C*. Show that *B* divides *AC* according to the golden section.

SOLUTION:

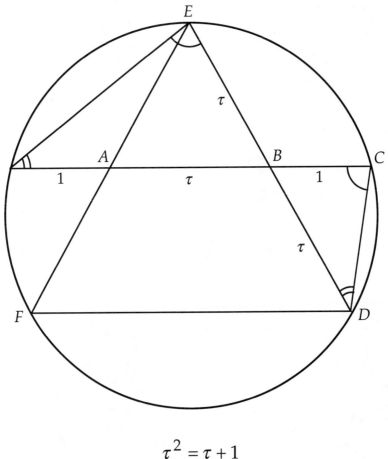

$$\tau^2 = \tau + 1$$

—Jan van de Craats

Tiling with Squares and Parallelograms

If squares are constructed externally on the sides of a parallelogram, their centers form a square.

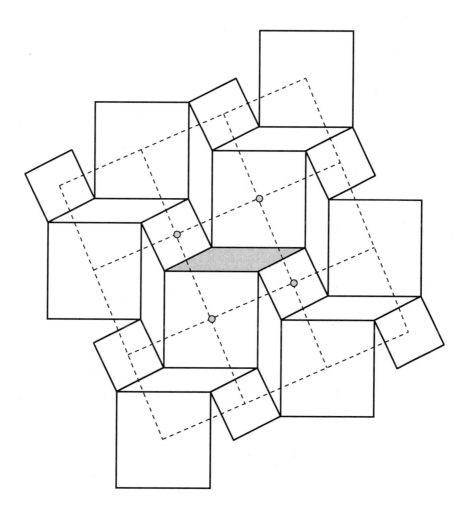

—Alfinio Flores

The Area of a Quadrilateral I

The area of a quadrilateral is less than or equal to half the product of the lengths of its diagonals, with equality if and only if the diagonals are perpendicular.

I. Convex quadrilaterals

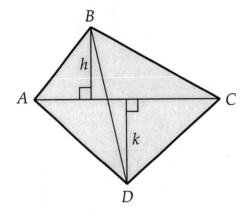

$$\text{Area} = \frac{1}{2}\overline{AC}\cdot(h+k)$$

$$\leq \frac{1}{2}\overline{AC}\cdot\overline{BD}$$

II. Concave quadrilaterals

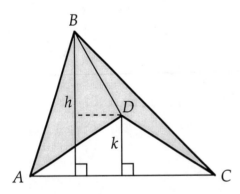

$$\text{Area} = \frac{1}{2}\overline{AC}\cdot(h-k)$$

$$\leq \frac{1}{2}\overline{AC}\cdot\overline{BD}$$

—David B. Sher, Ronald Skurnick,
and Dean C. Nataro

The Area of a Quadrilateral II

The area of a quadrilateral Q is equal to one-half the area of a parallelogram P whose sides are parallel to and equal in length to the diagonals of Q.

I. Q convex

Q:

P:

II. Q concave

Q:

P:

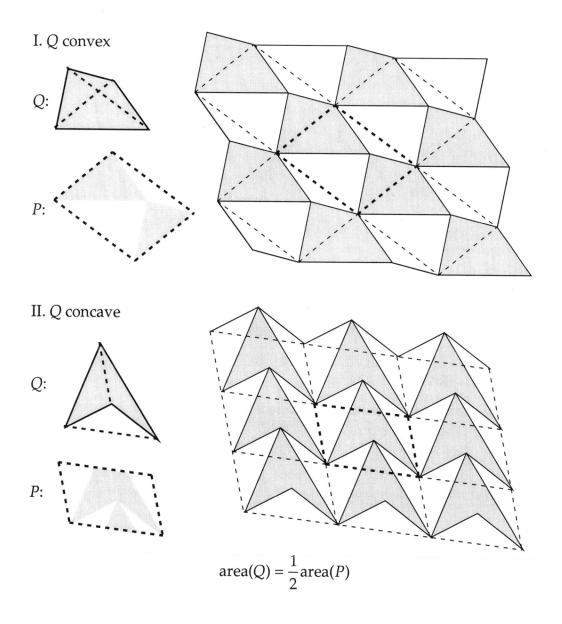

$$\text{area}(Q) = \frac{1}{2}\text{area}(P)$$

A Square Within a Square

If lines from the vertices of a square are drawn to the mid-points of adjacent sides (as shown in the figure), then the area of the smaller square so produced is one-fifth that of the given square.

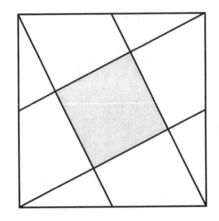

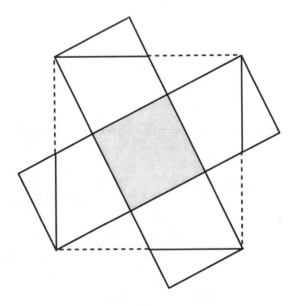

Areas and Perimeters of Regular Polygons

The area of a regular $2n$-gon inscribed in a circle is equal to one-half the radius of the circle times the perimeter of a regular n-gon similarly inscribed ($n \geq 3$).

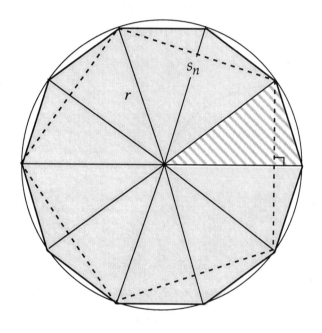

$$\frac{1}{2n}\text{area}(P_{2n}) = \frac{1}{2} \cdot r \cdot \frac{1}{2}s_n$$

$$\therefore \text{area}(P_{2n}) = \frac{r}{2}ns_n$$

$$= \frac{r}{2}\text{perimeter}(P_n)$$

COROLLARY [Bhāskara, *Lilāvati* (India, 12th century A.D.)]: The area of a circle is equal to one-half the product of its radius and circumference.

The Area of a Putnam Octagon

(Problem B1, 39th Annual William Lowell Putnam Mathematical Competition, 1978)

Find the area of a convex octagon that is inscribed in a circle and has four consecutive sides of length 3 units and the remaining four sides of length 2 units. Give the answer in the form $r + s\sqrt{t}$ with r, s, and t positive integers.

SOLUTION:

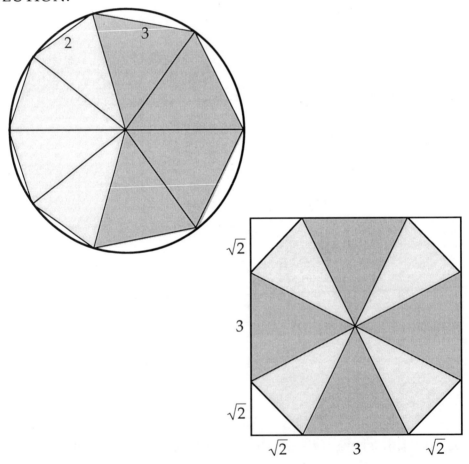

$$A = \left(3 + 2\sqrt{2}\right)^2 - 4 \cdot \frac{1}{2}\left(\sqrt{2}\right)^2$$

$$= 13 + 12\sqrt{2}$$

A Putnam Dodecagon

(Problem I-1, 24th Annual William Lowell Putnam Mathematical Competition, 1963)

(i) Show that a regular hexagon, six squares, and six equilateral triangles can be assembled without overlapping to form a regular dodecagon.

(ii) Let P_1, P_2, ..., P_{12} be the successive vertices of a regular dodecagon. Discuss the intersection(s) of the three diagonals P_1P_9, P_2P_{11}, and P_4P_{12}.

SOLUTION:

(i)

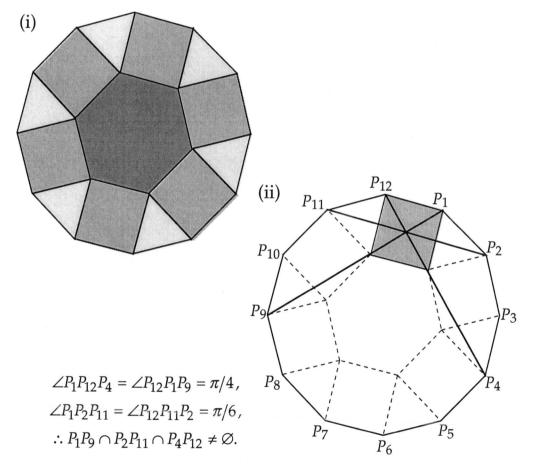

$$\angle P_1P_{12}P_4 = \angle P_{12}P_1P_9 = \pi/4,$$
$$\angle P_1P_2P_{11} = \angle P_{12}P_{11}P_2 = \pi/6,$$
$$\therefore P_1P_9 \cap P_2P_{11} \cap P_4P_{12} \neq \emptyset.$$

EXERCISE: Discuss the intersection(s) of the four diagonals P_1P_6, P_2P_9, P_3P_{11}, and P_4P_{12} (Problem F-4(b), *The AMATYC Review*, 1985, p. 61).

The Area of a Regular Dodecagon

A regular dodecagon with circumradius one has area three.

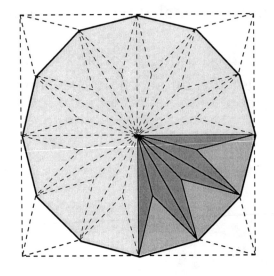

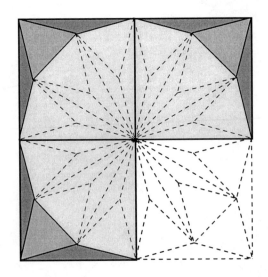

—J. Kürschák

Fair Allocation of a Pizza

THE PIZZA THEOREM: *If a pizza is divided into eight slices by making cuts at 45° angles from an arbitrary point in the pizza, then the sums of the areas of alternate slices are equal.*

PROOF:

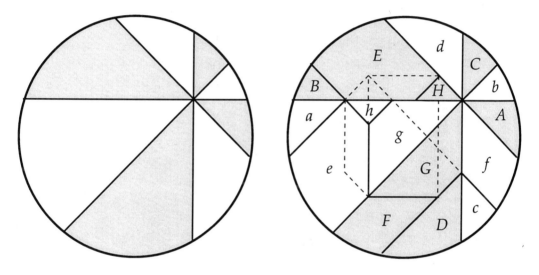

NOTES: This result, discovered by L. J. Upton, is true when n, the number of pieces, is 8, 12, 16, ..., but false for $n = 2, 4, 6, 10, 14, 18, ...$. The positive results are in the references. For the negative, the case $n = 4$ is easily handled, while if $n \equiv 2 \pmod 4$ we have the following argument of Don Coppersmith (IBM). It suffices, by continuity, to take the special point on the boundary of the unit circle and one of the chords to be a tangent at the point. Then the gray area can be expressed in terms of π and algebraic numbers in such a way that its equality with $\pi/2$ would yield an algebraic relationship for π, in contradiction to π's transcendence (details omitted).

REFERENCES

1. L. J. Upton, Problem 660, *Mathematics Magazine* 41 (1968) 46.
2. S. Rabinowitz, Problem 1325, *Crux Mathematicorum* 15 (1989) 120-122.

—Larry Carter and Stan Wagon

A Three-Circle Theorem

Given three nonintersecting mutually external circles, connect the intersection of internal common tangents of each pair of circles with the center of the other circle. Then the resulting three line segments are concurrent.

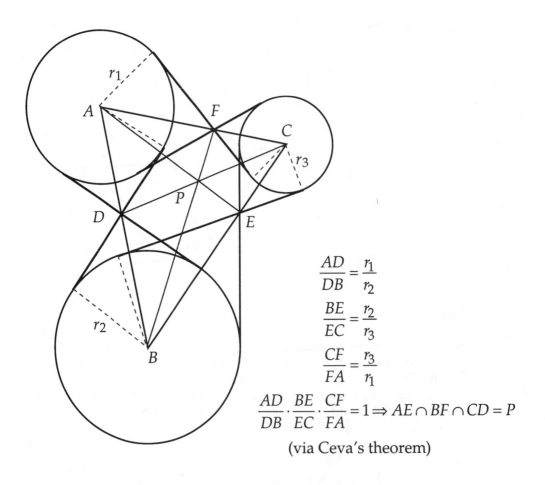

$$\frac{AD}{DB} = \frac{r_1}{r_2}$$

$$\frac{BE}{EC} = \frac{r_2}{r_3}$$

$$\frac{CF}{FA} = \frac{r_3}{r_1}$$

$$\frac{AD}{DB} \cdot \frac{BE}{EC} \cdot \frac{CF}{FA} = 1 \Rightarrow AE \cap BF \cap CD = P$$

(via Ceva's theorem)

—R. S. Hu

A Constant Chord

Suppose two circles Q and R intersect in A and B. A point P on the arc of Q which lies outside R is projected through A and B to determine chord CD of R. Prove that no matter where P is chosen on its arc, the length of the chord CD is always the same.

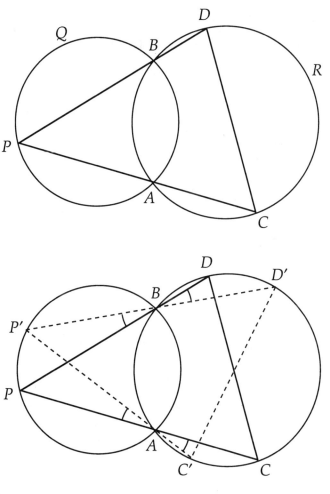

$$\angle C'AC = \angle P'AP = \angle P'BP = \angle D'BD$$

$$\overset{\frown}{C'C} = \overset{\frown}{D'D}, \quad \overset{\frown}{C'D'} = \overset{\frown}{CD}$$

$$C'D' = CD$$

A Putnam Area Problem

(Problem A2, 59th Annual William Lowell Putnam Mathematical Competition, 1998)

Let s be any arc of the unit circle lying entirely in the first quadrant. Let A be the area of the region lying below s and above the x-axis, and let B be the area of the region lying to the right of the y-axis and to the left of s. Prove that $A + B$ depends only on the arc length, and not the position, of s.

SOLUTION:

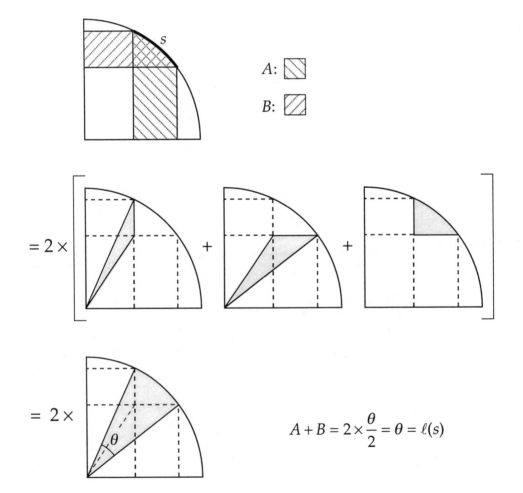

$$A + B = 2 \times \frac{\theta}{2} = \theta = \ell(s)$$

The Area Under a Polygonal Arch

The area under the polygonal arch generated by one vertex of a regular *n*-gon rolling along a straight line is three times the area of the polygon.

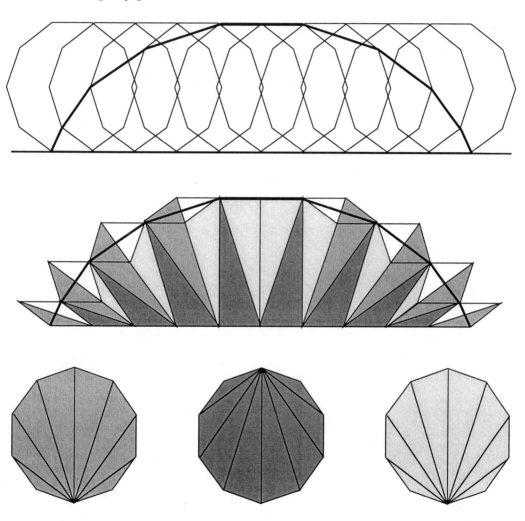

COROLLARY: The area under one arch of a cycloid is three times the area of the generating circle.

—Philip R. Mallinson

The Length of a Polygonal Arch

The length of the polygonal arch generated by one vertex of a regular n-gon rolling along a straight line is four times the length of the inradius plus four times the length of the circumradius of the n-gon.

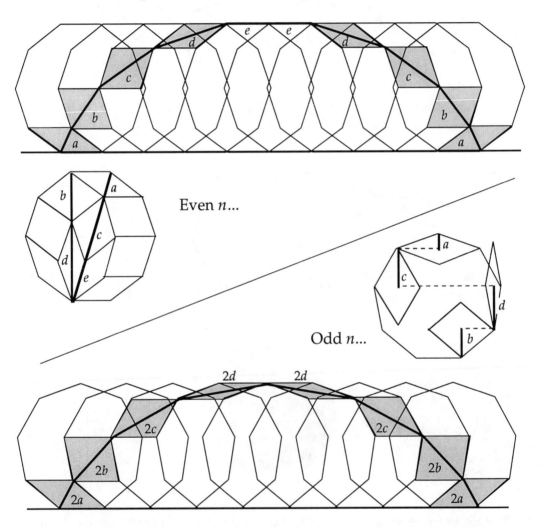

Even n...

Odd n...

COROLLARY: The arc length of one arch of a cycloid is eight times the radius of the generating circle.

— Philip R. Mallinson

The Volume of a Frustum of a Square Pyramid

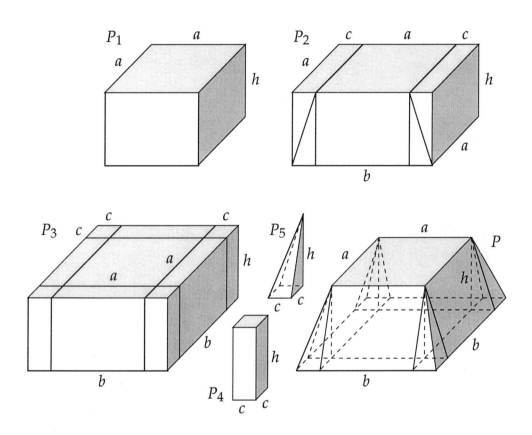

$$P_4 = 3P_5$$

$$P_1 + P_3 = 2P_2 + 4P_4 \Rightarrow P_1 + P_2 + P_3 = 3P_2 + 12P_5$$
$$= 3\left(P_2 + 4P_5\right) = 3P$$

$$\therefore V = \frac{h}{3}\left(a^2 + ab + b^2\right)$$

—Sidney J. Kung

The Product of Four (Positive) Numbers in Arithmetic Progression is Always the Difference of Two Squares

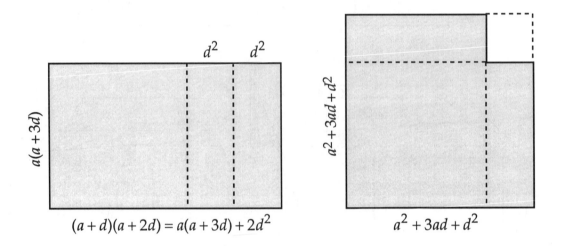

$$(a+d)(a+2d) = a(a+3d) + 2d^2 \qquad\qquad a^2 + 3ad + d^2$$

$$a(a+d)(a+2d)(a+3d) = (a^2 + 3ad + d^2)^2 - (d^2)^2$$

—RBN

Algebraic Areas III:
Factoring the Sum of Two Squares

$$x^2 + y^2 = \left(x + \sqrt{2xy} + y\right)\left(x - \sqrt{2xy} + y\right)$$

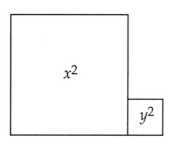

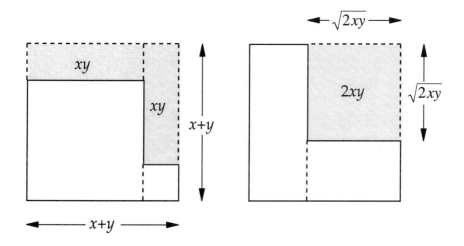

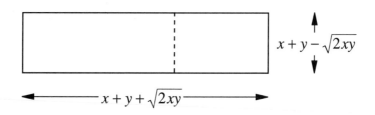

Trigonometry, Calculus, & Analytic Geometry

Sine of the Sum II

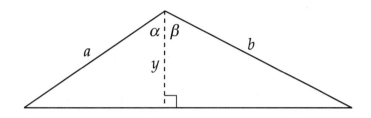

$$\alpha, \beta \in (0, \pi/2) \Rightarrow y = a\cos\alpha = b\cos\beta$$

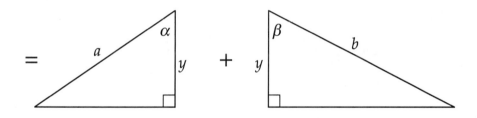

$$\frac{1}{2}ab\sin(\alpha + \beta) = \frac{1}{2}ay\sin\alpha + \frac{1}{2}by\sin\beta$$

$$= \frac{1}{2}ab\cos\beta\sin\alpha + \frac{1}{2}ba\cos\alpha\sin\beta$$

$$\therefore \sin(\alpha + \beta) = \sin\alpha\cos\beta + \cos\alpha\sin\beta$$

—Christopher Brueningsen

Sine of the Sum III

$$\sin(\alpha + \beta) = \sin\alpha\cos\beta + \sin\beta\cos\alpha$$

I.

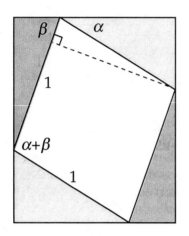 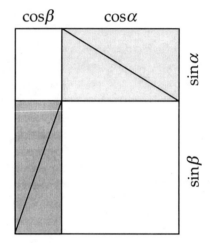

II.

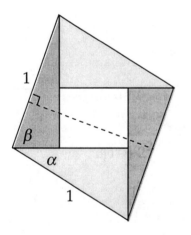 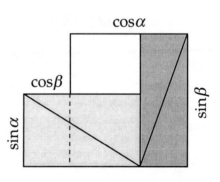

—Volker Priebe
and Edgar A. Ramos

Cosine of the Sum

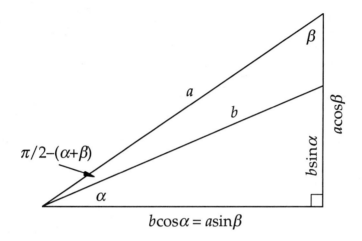

$$\frac{1}{2}ab\sin\left[\frac{\pi}{2}-(\alpha+\beta)\right]=\frac{1}{2}b\cos\alpha\cdot a\cos\beta-\frac{1}{2}a\sin\beta\cdot b\sin\alpha$$

$$\therefore \cos(\alpha+\beta)=\cos\alpha\cos\beta-\sin\alpha\sin\beta$$

—Sidney H. Kung

Geometry of Addition Formulas

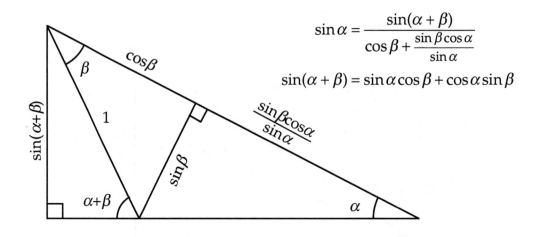

$$\sin\alpha = \frac{\sin(\alpha+\beta)}{\cos\beta + \dfrac{\sin\beta\cos\alpha}{\sin\alpha}}$$

$$\sin(\alpha+\beta) = \sin\alpha\cos\beta + \cos\alpha\sin\beta$$

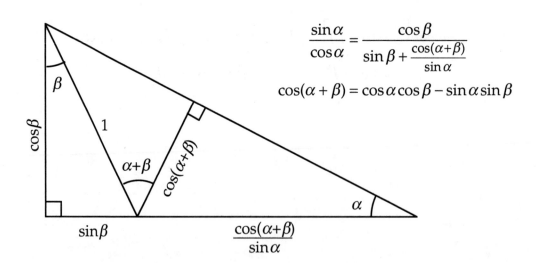

$$\frac{\sin\alpha}{\cos\alpha} = \frac{\cos\beta}{\sin\beta + \dfrac{\cos(\alpha+\beta)}{\sin\alpha}}$$

$$\cos(\alpha+\beta) = \cos\alpha\cos\beta - \sin\alpha\sin\beta$$

—Leonard M. Smiley

Geometry of Subtraction Formulas

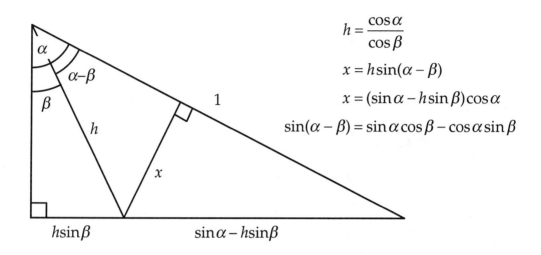

$$h = \frac{\cos\alpha}{\cos\beta}$$

$$x = h\sin(\alpha - \beta)$$

$$x = (\sin\alpha - h\sin\beta)\cos\alpha$$

$$\sin(\alpha - \beta) = \sin\alpha\cos\beta - \cos\alpha\sin\beta$$

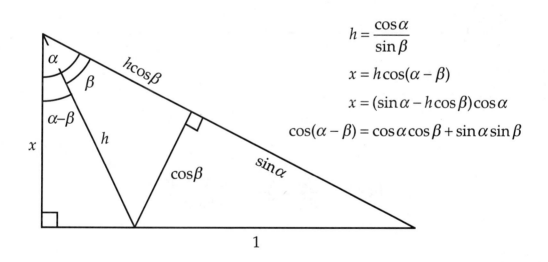

$$h = \frac{\cos\alpha}{\sin\beta}$$

$$x = h\cos(\alpha - \beta)$$

$$x = (\sin\alpha - h\cos\beta)\cos\alpha$$

$$\cos(\alpha - \beta) = \cos\alpha\cos\beta + \sin\alpha\sin\beta$$

—Leonard M. Smiley

The Difference Identity for Tangents I

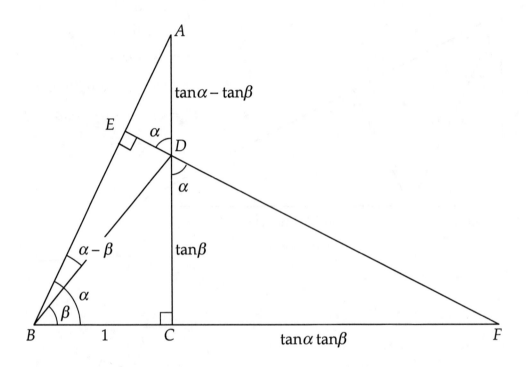

$$\frac{BF}{BE} = \frac{AD}{DE},$$

$$\therefore \tan(\alpha - \beta) = \frac{DE}{BE} = \frac{AD}{BF} = \frac{\tan\alpha - \tan\beta}{1 + \tan\alpha\tan\beta}.$$

—Guanshen Ren

The Difference Identity for Tangents II

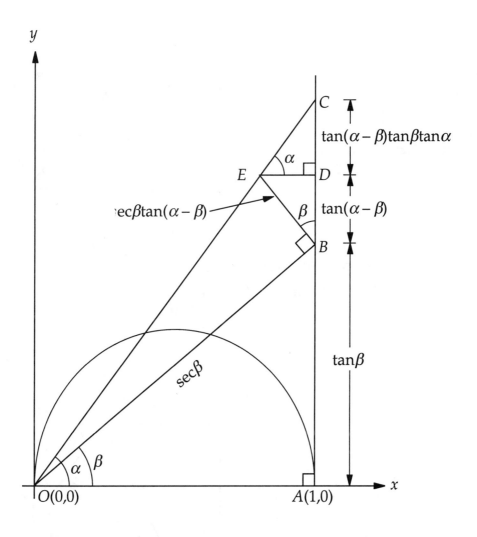

$$AC - AB = BD + DC$$

$$\therefore \tan\alpha - \tan\beta = \tan(\alpha - \beta) + \tan\alpha\tan\beta\tan(\alpha - \beta)$$

$$\tan(\alpha - \beta) = \frac{\tan\alpha - \tan\beta}{1 + \tan\alpha\tan\beta}.$$

—Fukuzo Suzuki

One Figure, Six Identities

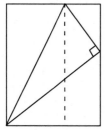

The figure

$$\sin(\alpha + \beta) = \sin\alpha\cos\beta + \cos\alpha\sin\beta$$
$$\cos(\alpha + \beta) = \cos\alpha\cos\beta - \sin\alpha\sin\beta$$

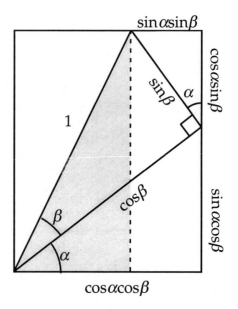

$$\sin(\alpha - \beta) = \sin\alpha\cos\beta - \cos\alpha\sin\beta$$
$$\cos(\alpha - \beta) = \cos\alpha\cos\beta + \sin\alpha\sin\beta$$

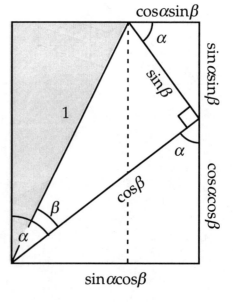

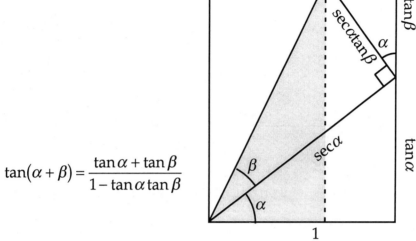

$$\tan(\alpha + \beta) = \frac{\tan\alpha + \tan\beta}{1 - \tan\alpha\tan\beta}$$

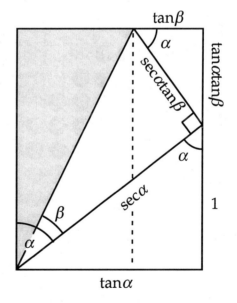

$$\tan(\alpha - \beta) = \frac{\tan\alpha - \tan\beta}{1 + \tan\alpha\tan\beta}$$

—RBN

The Double-Angle Formulas II

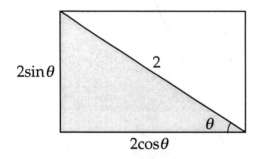

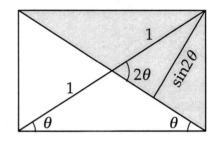

$$2\sin\theta\cos\theta = \sin 2\theta$$

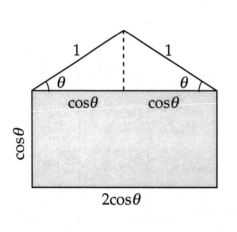

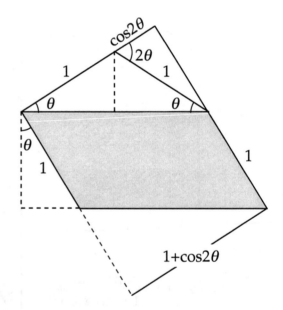

$$2\cos^2\theta = 1 + \cos 2\theta$$

—Yihnan David Gau

The Double-Angle Formulas III
(via the Laws of Sines and Cosines)

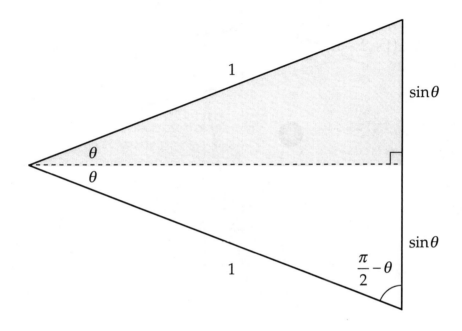

$$\frac{\sin 2\theta}{2\sin\theta} = \frac{\sin(\pi/2 - \theta)}{1} = \cos\theta$$

$$\sin 2\theta = 2\sin\theta\cos\theta$$

$$(2\sin\theta)^2 = 1^2 + 1^2 - 2\cdot 1\cdot 1\cdot\cos 2\theta$$

$$\cos 2\theta = 1 - 2\sin^2\theta$$

—Sidney H. Kung

The Sum-to-Product Identities I

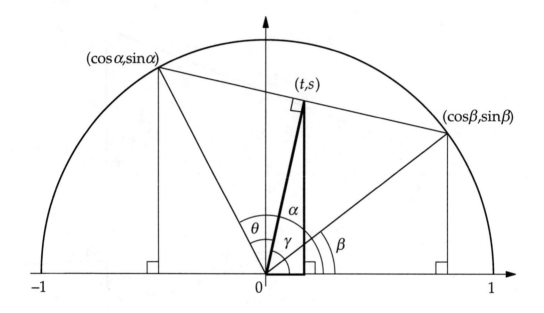

$$\theta = \frac{\alpha - \beta}{2}, \quad \gamma = \frac{\alpha + \beta}{2}$$

$$\frac{\sin \alpha + \sin \beta}{2} = s = \cos \frac{\alpha - \beta}{2} \sin \frac{\alpha + \beta}{2}$$

$$\frac{\cos \alpha + \cos \beta}{2} = t = \cos \frac{\alpha - \beta}{2} \cos \frac{\alpha + \beta}{2}$$

—Sidney H. Kung

The Sum-to-Product Identities II

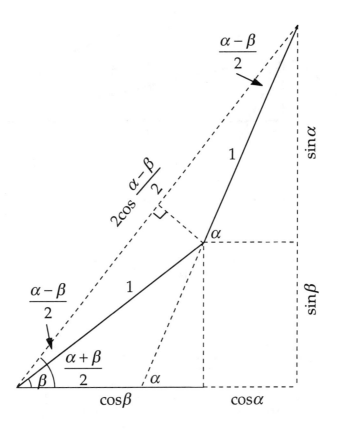

$$\cos\alpha + \cos\beta = 2\cos\frac{\alpha-\beta}{2}\cos\frac{\alpha+\beta}{2}$$

$$\sin\alpha + \sin\beta = 2\cos\frac{\alpha-\beta}{2}\sin\frac{\alpha+\beta}{2}$$

—Yokio Kobayashi

The Difference-to-Product Identities I

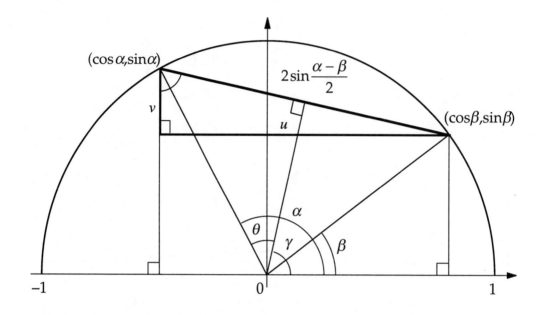

$$\theta = \frac{\alpha - \beta}{2}, \quad \gamma = \frac{\alpha + \beta}{2}$$

$$\sin\alpha - \sin\beta = v = 2\sin\frac{\alpha - \beta}{2}\cos\frac{\alpha + \beta}{2}$$

$$\cos\beta - \cos\alpha = u = 2\sin\frac{\alpha - \beta}{2}\sin\frac{\alpha + \beta}{2}$$

— Sidney H. Kung

The Difference-to-Product Identities II

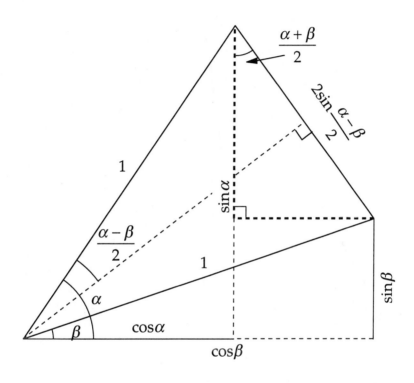

$$\cos \beta - \cos \alpha = 2\sin\frac{\alpha - \beta}{2}\sin\frac{\alpha + \beta}{2}$$

$$\sin \alpha - \sin \beta = 2\sin\frac{\alpha - \beta}{2}\cos\frac{\alpha + \beta}{2}$$

— Yokio Kobayashi

Adding Like Sines

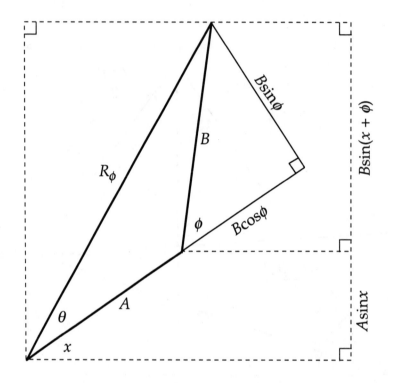

$$R_\phi = \sqrt{A^2 + B^2 + 2AB\cos\phi}, \quad \tan\theta = \frac{B\sin\phi}{A + B\cos\phi}$$

$$A\sin x + B\sin(x + \phi) = R_\phi \sin(x + \theta)$$

$$\phi = \frac{\pi}{2} \implies \tan\theta = \frac{B}{A}$$

$$\therefore A\sin x + B\cos x = \sqrt{A^2 + B^2}\,\sin(x + \theta)$$

—Rick Mabry
and Paul Deiermann

A Complex Approach to the Laws of Sines and Cosines

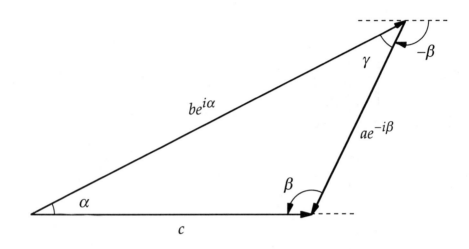

$$c = be^{i\alpha} + ae^{-i\beta} = (b\cos\alpha + a\cos\beta) + i(b\sin\alpha - a\sin\beta)$$

$$c \text{ real} \Rightarrow b\sin\alpha - a\sin\beta = 0 \Rightarrow \frac{a}{\sin\alpha} = \frac{b}{\sin\beta}$$

$$c^2 = |c|^2 = (b\cos\alpha + a\cos\beta)^2 + (b\sin\alpha - a\sin\beta)^2$$
$$= a^2 + b^2 + 2ab\cos(\alpha + \beta)$$
$$= a^2 + b^2 - 2ab\cos\gamma$$

—William V. Grounds

Eisenstein's Duplication Formula

(G. Eisenstein, *Mathematische Werke*, Chelsea, New York, 1975, p. 411)

$$2\csc\theta = \tan\theta + \cot\theta$$

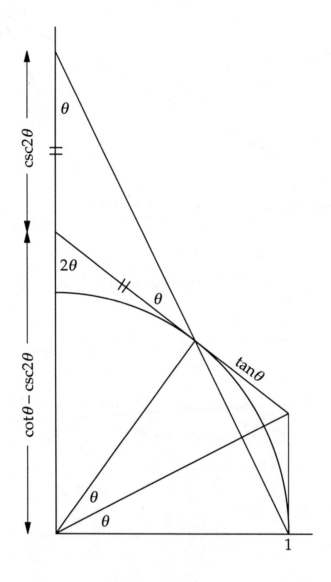

—Lin Tan

A Familiar Limit for e

$$\lim_{n\to\infty}\left(1+\frac{1}{n}\right)^n = e$$

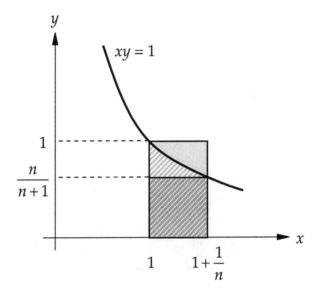

$$\frac{1}{n}\cdot\frac{n}{n+1} \le \ln\left(1+\frac{1}{n}\right) \le \frac{1}{n}\cdot 1$$

$$\frac{n}{n+1} \le n\cdot\ln\left(1+\frac{1}{n}\right) \le 1$$

$$\therefore \lim_{n\to\infty}\ln\left(1+\frac{1}{n}\right)^n = 1$$

A Common Limit

$$\lim_{x \to \infty} \frac{x}{e^x} = 0$$

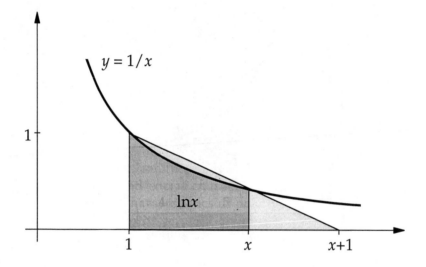

$$\ln x < \frac{1}{2} x$$

$$\therefore \lim_{x \to \infty} \frac{x}{e^x} = \lim_{x \to \infty} \frac{1}{e^{x - \ln x}} = 0$$

—Alan H. Stein
and Dennis McGavran

Geometric Evaluation of a Limit

$$\sqrt{2+\sqrt{2+\sqrt{2+\sqrt{\cdots}}}} = 2$$

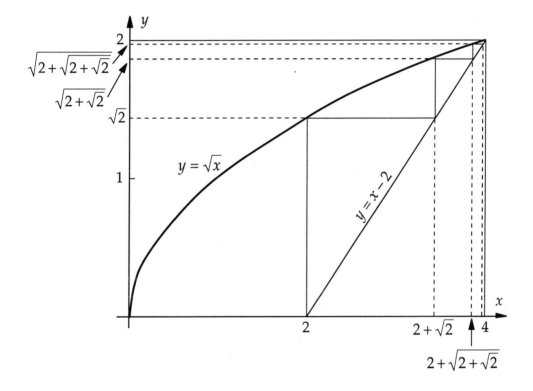

—Guanshen Ren

The Derivative of the Inverse Sine

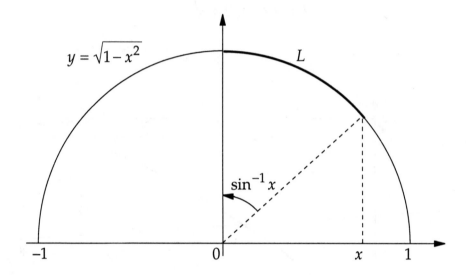

$$L = \sin^{-1} x = \int_0^x \frac{1}{\sqrt{1-t^2}}\, dt$$

$$\therefore \frac{d}{dx} \sin^{-1} x = \frac{1}{\sqrt{1-x^2}}$$

—Craig Johnson

The Logarithm of a Product

$$\ln ab = \ln a + \ln b$$

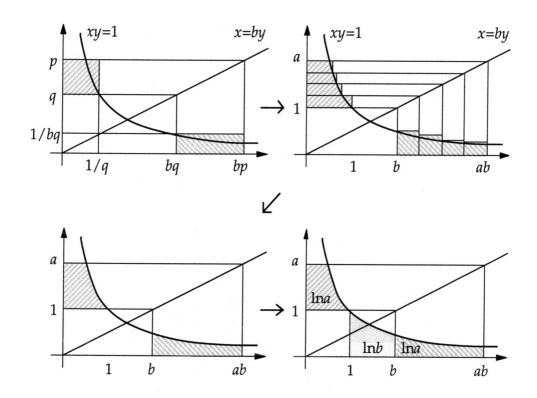

$$\text{Area}(\boxed{}) = \text{Area}(\boxed{})$$

—Jeffrey Ely

An Integral of a Sum of Reciprocal Powers

$$\int_0^1 \left(t^{p/q} + t^{q/p}\right) dt = 1$$

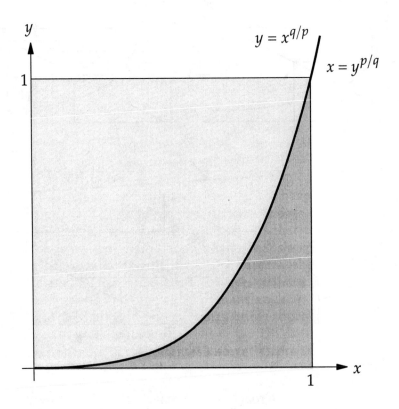

$y = x^{q/p}$

$x = y^{p/q}$

—Peter R. Newbury

The Arctangent Integral

$$\arctan x = \int_0^x \frac{1}{1+t^2}\, dt$$

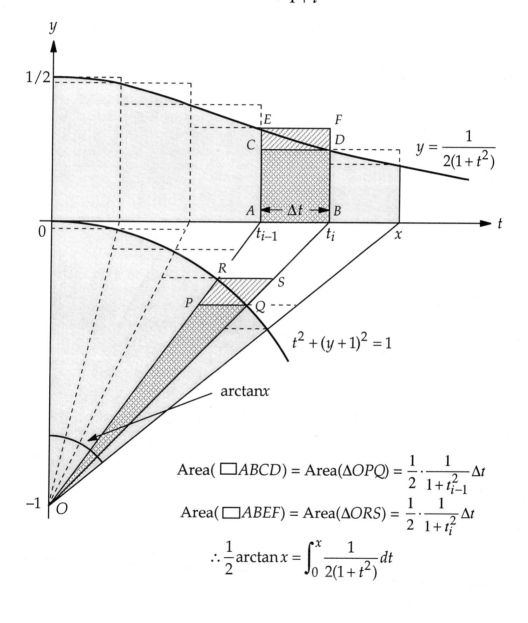

$$\text{Area}(\square ABCD) = \text{Area}(\triangle OPQ) = \frac{1}{2}\cdot\frac{1}{1+t_{i-1}^2}\Delta t$$

$$\text{Area}(\square ABEF) = \text{Area}(\triangle ORS) = \frac{1}{2}\cdot\frac{1}{1+t_i^2}\Delta t$$

$$\therefore \frac{1}{2}\arctan x = \int_0^x \frac{1}{2(1+t^2)}\, dt$$

—Aage Bondesen

The Method of Last Resort
(Weierstrass Substitution)

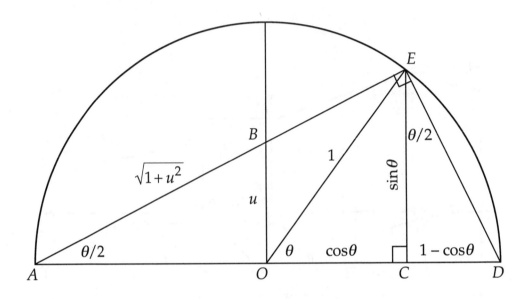

$$u = \tan\frac{\theta}{2}, \quad \overline{DE} = 2\sin\frac{\theta}{2} = \frac{2u}{\sqrt{1+u^2}}$$

$$\frac{\overline{CE}}{\overline{DE}} = \frac{\overline{OA}}{\overline{BA}} \quad \Rightarrow \quad \sin\theta = \frac{2u}{1+u^2}$$

$$\frac{\overline{CD}}{\overline{DE}} = \frac{\overline{OB}}{\overline{BA}} \quad \Rightarrow \quad \cos\theta = \frac{1-u^2}{1+u^2}$$

—Paul Deiermann

The Trapezoidal Rule
(for Increasing Functions)

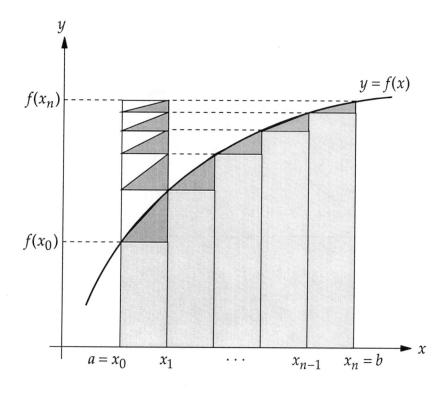

$$\int_a^b f(x)\,dx = \sum_{i=0}^{n-1} f(x_i)\frac{b-a}{n} + \frac{1}{2}[f(x_n) - f(x_0)]\frac{b-a}{n}$$

—Jesús Urías

Construction of a Hyperbola

I.

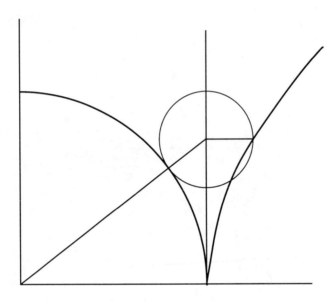

II.

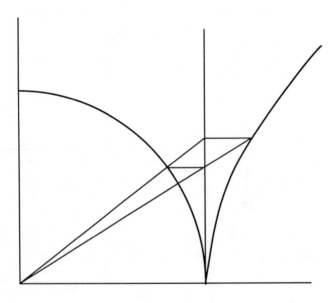

—Ernest J. Eckert

The Focus and Directrix of an Ellipse

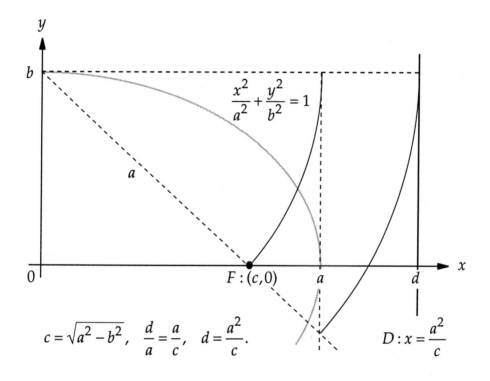

$$c = \sqrt{a^2 - b^2}, \quad \frac{d}{a} = \frac{a}{c}, \quad d = \frac{a^2}{c}.$$

—Michel Bataille

Inequalities

The Arithmetic Mean–Geometric Mean Inequality IV

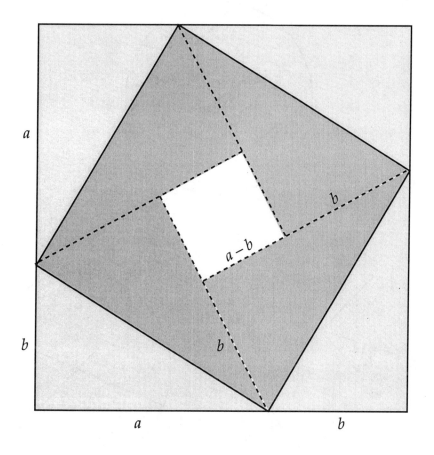

$$(a+b)^2 \geq 4ab \quad \Rightarrow \quad \frac{a+b}{2} \geq \sqrt{ab}$$

—Ayoub B. Ayoub

The Arithmetic Mean–Geometric Mean Inequality V

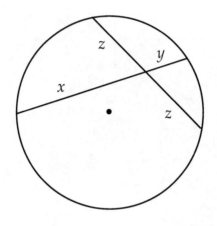

$$z^2 = xy$$

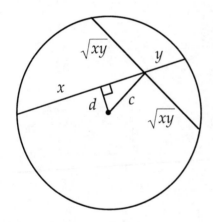

$$d < c \Rightarrow x + y > 2\sqrt{xy}$$

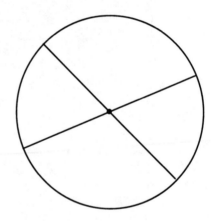

$$d = c = 0 \Rightarrow x + y = 2\sqrt{xy}$$

—Sidney H. Kung

The Arithmetic Mean–Geometric Mean Inequality VI

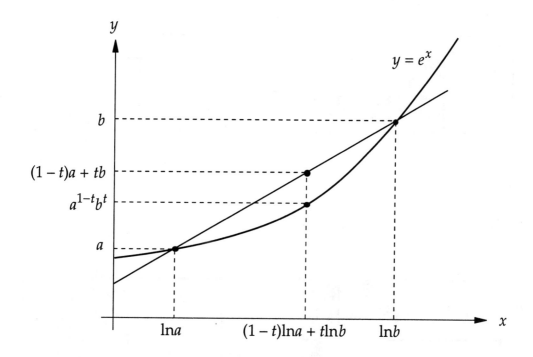

$$0 < a < b,\, 0 < t < 1 \Rightarrow (1-t)a + tb > a^{1-t}b^t$$

$$t = \frac{1}{2} \Rightarrow \frac{a+b}{2} > \sqrt{ab}$$

—Michael K. Brozinsky

The Arithmetic Mean–Geometric Mean Inequality for Three Positive Numbers

LEMMA: $ab + bc + ac \leq a^2 + b^2 + c^2$

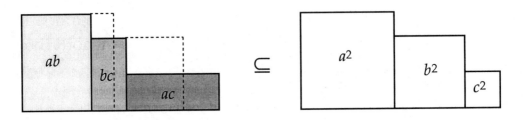

THEOREM: $3abc \leq a^3 + b^3 + c^3$

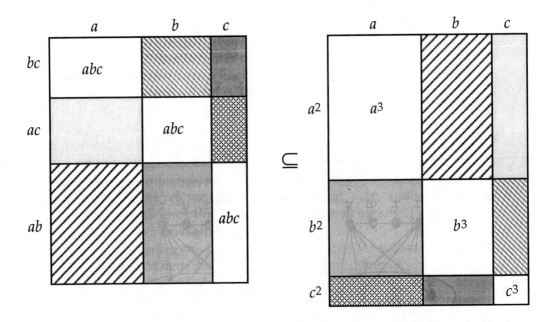

—Claudi Alsina

The Arithmetic–Geometric–Harmonic Mean Inequality

$$a, b > 0 \Rightarrow \frac{a+b}{2} \geq \sqrt{ab} \geq \frac{2ab}{a+b}$$

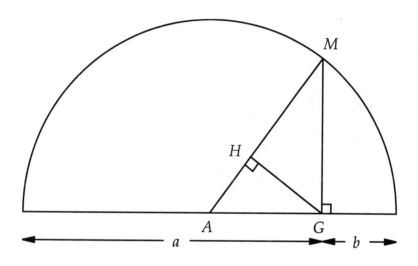

$$\overline{AM} = \frac{a+b}{2}, \ \overline{GM} = \sqrt{ab}, \ \overline{HM} = \frac{2ab}{a+b},$$
$$\overline{AM} \geq \overline{GM} \geq \overline{HM}.$$

—Pappus of Alexandria (circa A.D. 320)

The Arithmetic–Logarithmic–Geometric Mean Inequality

$$b > a > 0 \Rightarrow \frac{a+b}{2} > \frac{b-a}{\ln b - \ln a} > \sqrt{ab}$$

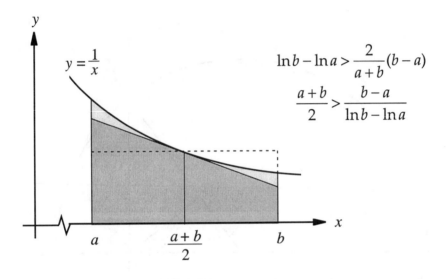

$$\ln b - \ln a > \frac{2}{a+b}(b-a)$$

$$\frac{a+b}{2} > \frac{b-a}{\ln b - \ln a}$$

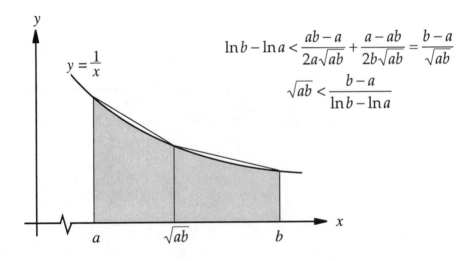

$$\ln b - \ln a < \frac{ab - a}{2a\sqrt{ab}} + \frac{a - ab}{2b\sqrt{ab}} = \frac{b-a}{\sqrt{ab}}$$

$$\sqrt{ab} < \frac{b-a}{\ln b - \ln a}$$

—RBN

The Mean of the Squares Exceeds the Square of the Mean

$$\frac{1}{n}\sum_{i=1}^{n}x_i^2 \geq \left(\frac{1}{n}\sum_{i=1}^{n}x_i\right)^2$$

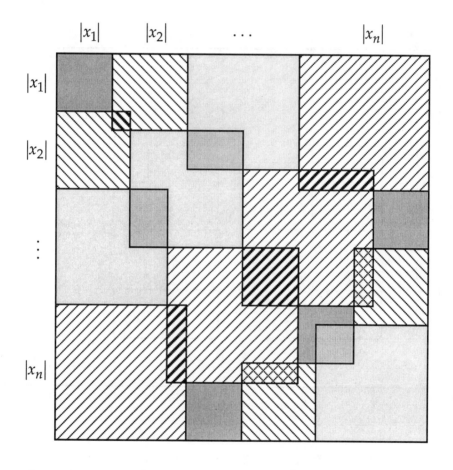

$$n\left(x_1^2 + x_2^2 + \cdots + x_n^2\right) \geq \left(|x_1| + |x_2| + \cdots + |x_n|\right)^2 \geq (x_1 + x_2 + \cdots + x_n)^2$$

$$\therefore \frac{x_1^2 + x_2^2 + \cdots + x_n^2}{n} \geq \left(\frac{x_1 + x_2 + \cdots + x_n}{n}\right)^2$$

—RBN

The Chebyshev Inequality for Positive Monotone Sequences

$$\sum_{i=1}^{n} x_i \sum_{i=1}^{n} y_i \leq n \sum_{i=1}^{n} x_i y_i$$

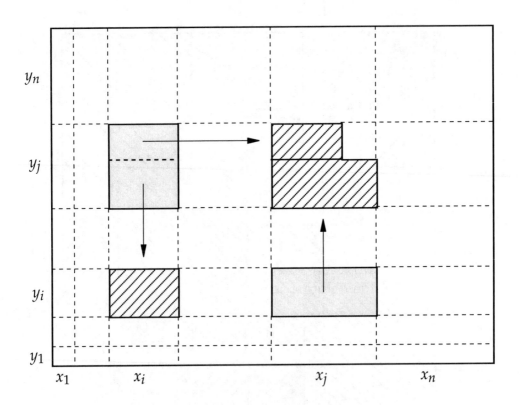

$$x_i < x_j \ \& \ y_i < y_j \Rightarrow x_i y_j + x_j y_i \leq x_i y_i + x_j y_j$$

$$\therefore (x_1 + x_2 + \cdots + x_n)(y_1 + y_2 + \cdots + y_n) \leq n(x_1 y_1 + x_2 y_2 + \cdots + x_n y_n)$$

—RBN

Jordan's Inequality

$$0 \le x \le \frac{\pi}{2} \Rightarrow \frac{2x}{\pi} \le \sin x \le x$$

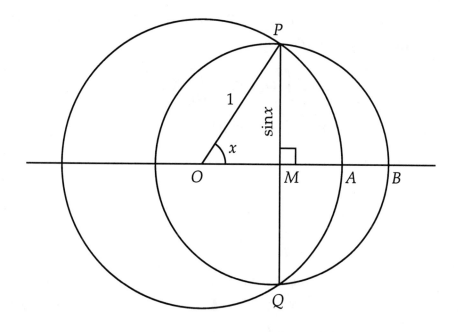

$$OB = OM + MP \ge OA \Rightarrow \overset{\frown}{PBQ} \ge \overset{\frown}{PAQ} \ge \overline{PQ}$$

$$\Rightarrow \pi \sin x \ge 2x \ge 2\sin x$$

$$\Rightarrow \frac{2x}{\pi} \le \sin x \le x$$

—Feng Yuefeng

Young's Inequality

(W. H. Young, "On classes of summable functions and their Fourier series," *Proc. Royal Soc.* (A), 87 (1912) 225-229)

THEOREM: *Let φ and ψ be two functions, continuous, vanishing at the origin, strictly increasing, and inverse to each other. Then for $a,b \geq 0$ we have*

$$ab \leq \int_0^a \varphi(x)dx + \int_0^b \psi(y)dy$$

with equality if and only if $b = \varphi(a)$.

PROOF:

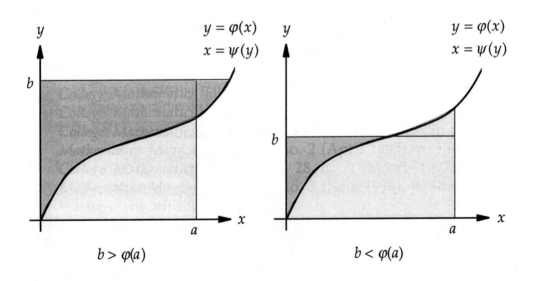

Integer Sums

Sums of Integers III

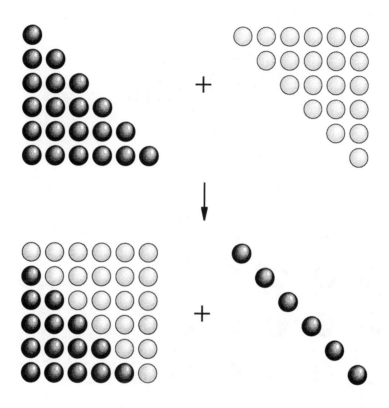

$$1 + 2 + \cdots + n = \frac{1}{2}\left(n^2 + n\right)$$

—S. J. Farlow

Sums of Consecutive Positive Integers

Every integer $N > 1$, not a power of two, can be expressed as the sum of two or more consecutive positive integers.

$$N = 2^n(2k+1) \quad (n \geq 0, k \geq 1),$$
$$m = \min\{2^{n+1}, 2k+1\},$$
$$M = \max\{2^{n+1}, 2k+1\},$$
$$2N = mM.$$

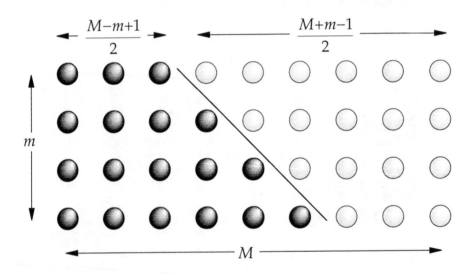

$$N = \left(\frac{M-m+1}{2}\right) + \left(\frac{M-m+1}{2}+1\right) + \cdots + \left(\frac{M+m-1}{2}\right).$$

REFERENCES

1. P. Ross, Problem 1358, *Mathematics Magazine* 63 (1990), 350.

2. J. V. Wales, Jr., Solution to Problem 1358, *Mathematics Magazine* 64 (1991), 351.

—C. L. Frenzen

Consecutive Sums of Consecutive Integers II

$$T_k = 1 + 2 + \cdots + k \Rightarrow$$

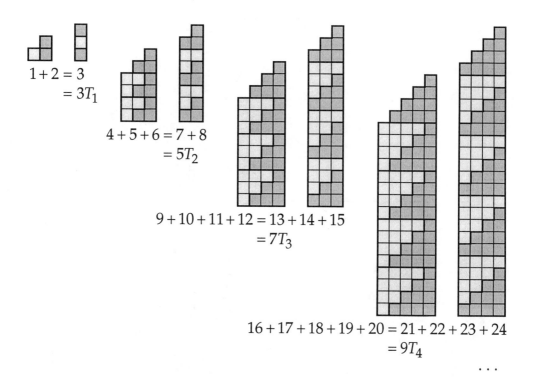

$$1 + 2 = 3$$
$$= 3T_1$$

$$4 + 5 + 6 = 7 + 8$$
$$= 5T_2$$

$$9 + 10 + 11 + 12 = 13 + 14 + 15$$
$$= 7T_3$$

$$16 + 17 + 18 + 19 + 20 = 21 + 22 + 23 + 24$$
$$= 9T_4$$

$$\cdots$$

$$n^2 + (n^2 + 1) + \cdots + (n^2 + n) = (n^2 + n + 1) + \cdots + (n^2 + 2n)$$
$$= (2n + 1)T_n$$

Sums of Squares VI

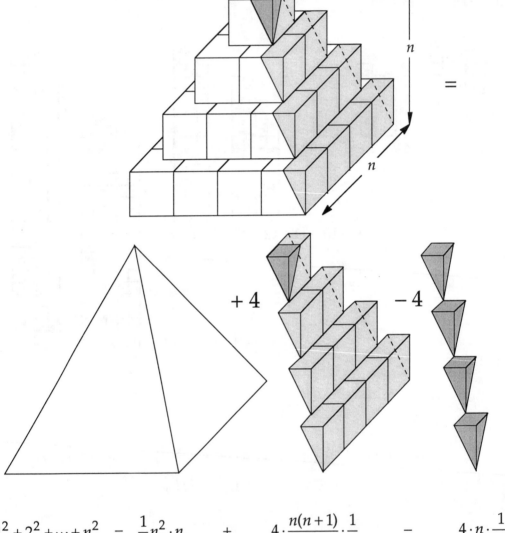

$$1^2 + 2^2 + \cdots + n^2 \; = \; \frac{1}{3}n^2 \cdot n \quad + \quad 4 \cdot \frac{n(n+1)}{2} \cdot \frac{1}{4} \quad - \quad 4 \cdot n \cdot \frac{1}{12}$$

$$= \; \frac{1}{6}n(n+1)(2n+1).$$

—I. A. Sakmar

Sums of Squares VII

$$\sum_{k=1}^{n} k^2 = \frac{n(n+1)(2n+1)}{6}$$

1^2

2^2

3^2

4^2

$\sum k^2$

$\sum k^2$

$\sum k^2$

$3\sum k^2$

$$6\sum k^2 = n(n+1)(2n+1)$$

—Nanny Wermuth
and Hans-Jürgen Schuh

Sums of Squares VIII

$$k^2 = 1 + 3 + \cdots + (2k-1) \Rightarrow \sum_{k=1}^{n} k^2 = \frac{n(n+1)(2n+1)}{6}$$

```
          1                         1                      2n–1
        1   3                     3   1                2n–3  2n–3
      1   3   5          +      5   3   1      +     2n–5  2n–5  2n–5
        ⋮                        ⋮                       ⋮
  1  3  5  ⋯  2n–3        2n–3  ⋯  5  3  1        3  3   ⋯   3  3
1  3  5  ⋯  2n–3 2n–1   2n–1 2n–3  ⋯  5  3  1   1  1  1  ⋯  1  1  1
```

```
                    2n+1
                 2n+1  2n+1
     =         2n+1  ⋯  2n+1
                    ⋮
             2n+1 2n+1  ⋯  2n+1
           2n+1 2n+1   ⋯      2n+1
```

$$3\left(1^2 + 2^2 + \cdots + n^2\right) = (2n+1)(1 + 2 + \cdots + n)$$

$$\therefore 1^2 + 2^2 + \cdots + n^2 = \frac{2n+1}{3} \cdot \frac{n(n+1)}{2}$$

Sums of Squares IX (via Centroids)

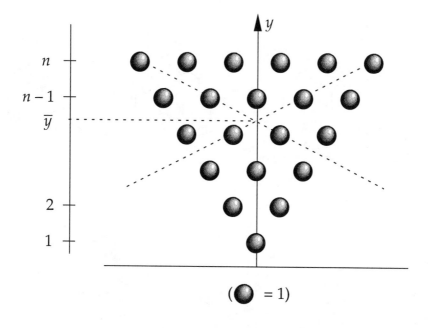

$$(\bullet = 1)$$

$$\bar{y} = 1 + \frac{2}{3}(n-1) = \frac{1 \cdot 1 + 2 \cdot 2 + \cdots + n \cdot n}{1 + 2 + \cdots + n}$$

$$\therefore 1^2 + 2^2 + \cdots + n^2 = \frac{n(n+1)}{2} \cdot \frac{1}{3}(2n+1) = \frac{1}{6}n(n+1)(2n+1)$$

—Sidney H. Kung

Sums of Odd Squares

$$1^2 + 3^2 + \cdots + (2n-1)^2 = \frac{n(2n-1)(2n+1)}{3}$$

1^2

3^2

$\vdots$

$(2n-1)^2$

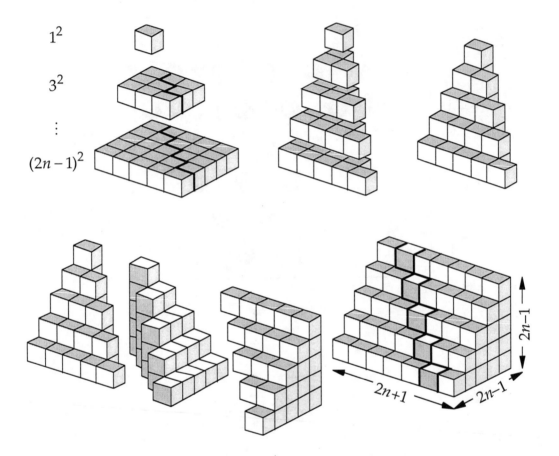

$$3 \times \left[1^2 + 3^2 + \cdots + (2n-1)^2\right] = \left[1 + 2 + \cdots + (2n-1)\right] \times (2n+1)$$

$$= \frac{(2n-1)(2n)(2n+1)}{2} = n(2n-1)(2n+1)$$

—RBN

Sums of Sums of Squares

$$\sum_{k=1}^{n}\sum_{i=1}^{k}i^2 = \frac{1}{3}\binom{n+1}{2}\binom{n+2}{2}$$

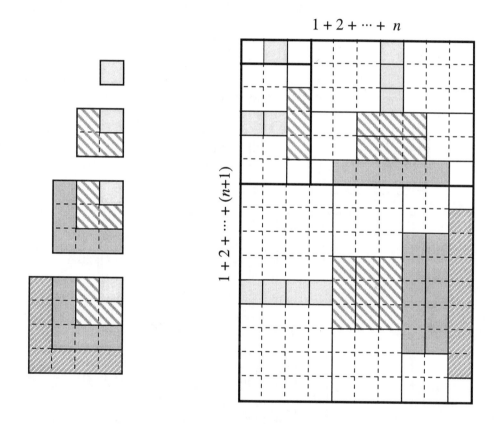

$$3\left(1^2\right)+3\left(1^2+2^2\right)+3\left(1^2+2^2+3^2\right)+\cdots+3\left(1^2+2^2+\cdots+n^2\right)=\binom{n+1}{2}\binom{n+2}{2}$$

—C. G. Wastun

Pythagorean Runs

$$3^2 + 4^2 = 5^2$$

$$10^2 + 11^2 + 12^2 = 13^2 + 14^2$$

$$21^2 + 22^2 + 23^2 + 24^2 = 25^2 + 26^2 + 27^2$$

$$\vdots$$

$$T_n = 1 + 2 + \cdots + n \Rightarrow (4T_n - n)^2 + \cdots + (4T_n)^2 = (4T_n + 1)^2 + \cdots + (4T_n + n)^2$$

e.g., $n = 3$:

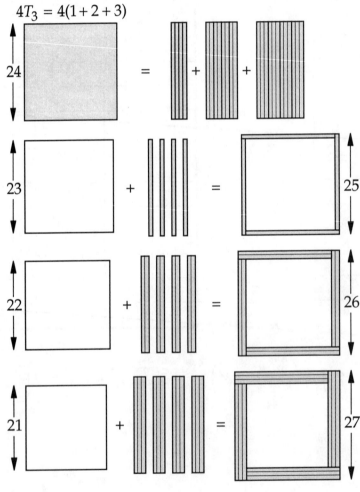

—Michael Boardman

Sums of Cubes VII

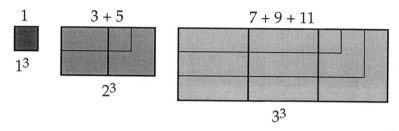

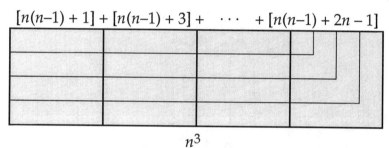

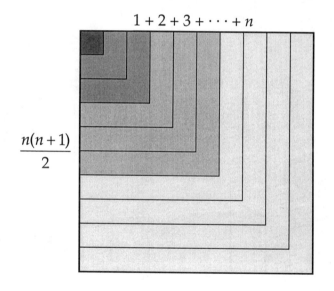

$$1^3 + 2^3 + \cdots + n^3 = 1 + 3 + 5 + \cdots + 2\frac{n(n+1)}{2} - 1 = \left[\frac{n(n+1)}{2}\right]^2$$

—Alfinio Flores

Sums of Integers as Sums of Cubes

$$2+3+4 = 1+8$$
$$5+6+7+8+9 = 8+27$$
$$10+11+12+13+14+15+16 = 27+64$$
$$\vdots$$
$$(n^2+1)+(n^2+2)+\cdots+(n+1)^2 = n^3+(n+1)^3$$

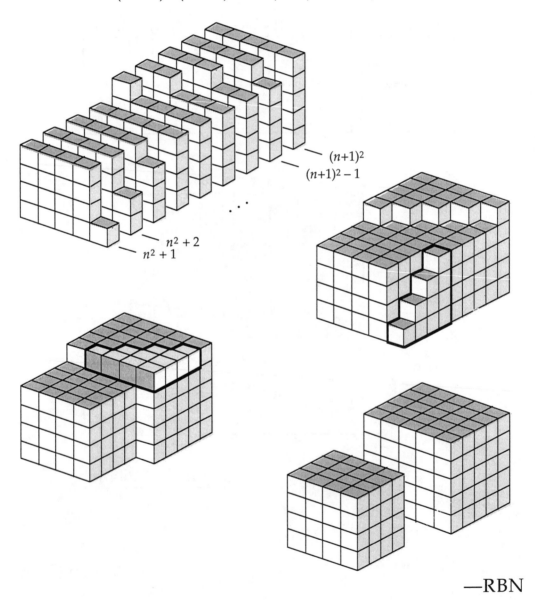

—RBN

The Square of Any Odd Number is the Difference Between Two Triangular Numbers

$$1 + 2 + \cdots + k = T_k \Rightarrow (2n+1)^2 = T_{3n+1} - T_n$$

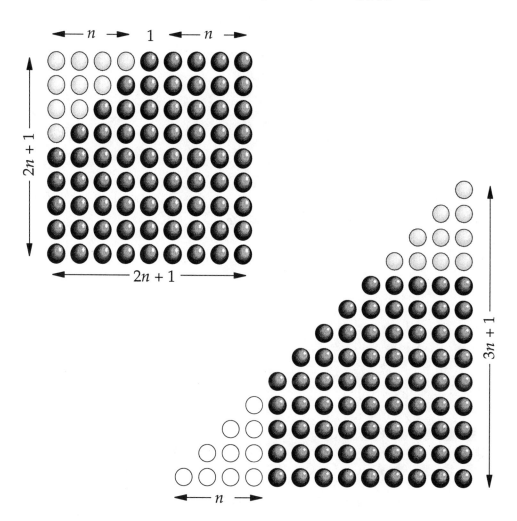

—RBN

Triangular Numbers Mod 3

$$t_n = 1 + 2 + \cdots + n \Longrightarrow \begin{cases} t_n \equiv 1 \bmod 3, & n \equiv 1 \bmod 3 \\ t_n \equiv 0 \bmod 3, & n \not\equiv 1 \bmod 3 \end{cases}$$

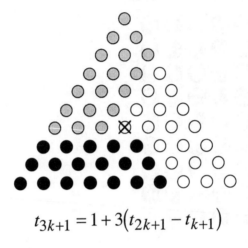

$$t_{3k+1} = 1 + 3(t_{2k+1} - t_{k+1})$$

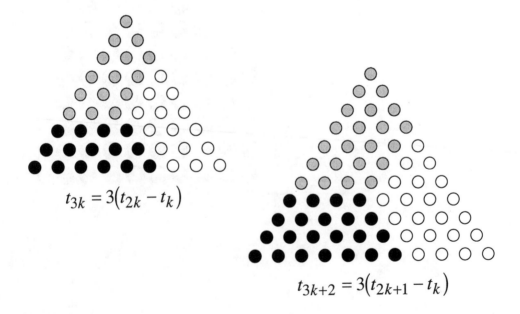

$$t_{3k} = 3(t_{2k} - t_k)$$

$$t_{3k+2} = 3(t_{2k+1} - t_k)$$

Sums of Triangular Numbers IV: Counting Cannonballs

$$T_k = 1 + 2 + \cdots + k \Rightarrow \sum_{k=1}^{n} T_k = \sum_{k=1}^{n} k(n-k+1)$$

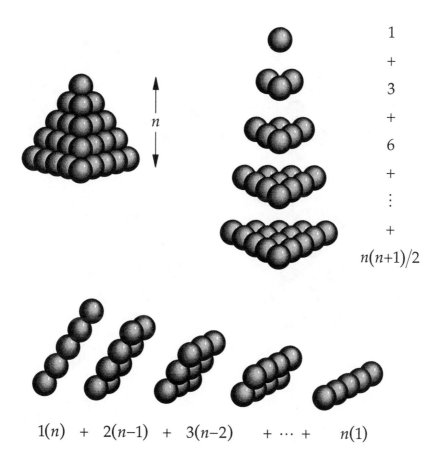

1

+

3

+

6

+

$\vdots$

+

$n(n+1)/2$

$1(n) \quad + \quad 2(n{-}1) \quad + \quad 3(n{-}2) \quad + \cdots + \quad n(1)$

—Deanna B. Haunsperger
and Stephen F. Kennedy

Alternating Sums of Triangular Numbers

$$T_k = 1 + 2 + \cdots + k \Rightarrow \sum_{k=1}^{2n-1} (-1)^{k+1} T_k = n^2$$

—RBN

The Sum of the Squares of Consecutive Triangular Numbers is Triangular

$$T_n = 1 + 2 + \cdots + n \Rightarrow T_{n-1}^2 + T_n^2 = T_{n^2}$$

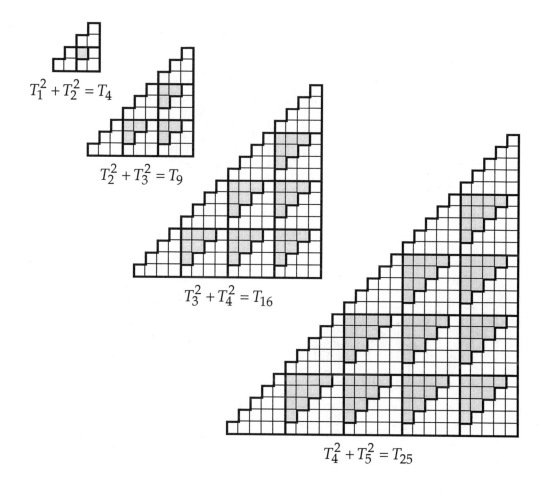

$T_1^2 + T_2^2 = T_4$

$T_2^2 + T_3^2 = T_9$

$T_3^2 + T_4^2 = T_{16}$

$T_4^2 + T_5^2 = T_{25}$

NOTE:

This is a companion result to the more familiar $T_{n-1} + T_n = n^2$:

—RBN

Recursion for Triangular Numbers

$$T_k = 1 + 2 + \cdots + k \Rightarrow T_{n+1} = \frac{n+2}{n} T_n$$

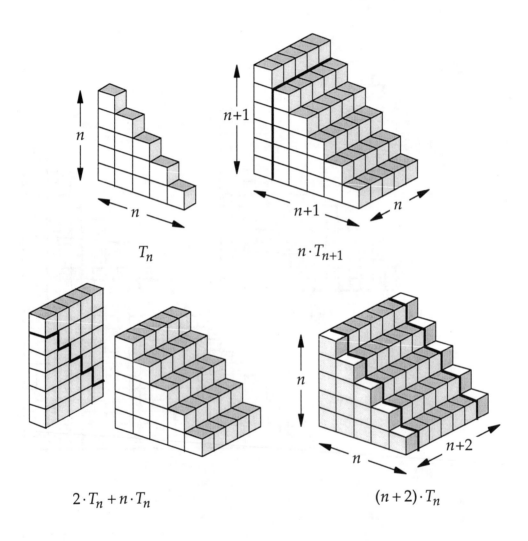

T_n

$n \cdot T_{n+1}$

$2 \cdot T_n + n \cdot T_n$

$(n+2) \cdot T_n$

$$n \cdot T_{n+1} = (n+2) \cdot T_n \Rightarrow T_{n+1} = \frac{n+2}{n} T_n$$

Identities for Triangular Numbers II

$$T_n = 1 + 2 + \cdots + n \Rightarrow$$

$$T_n T_k + T_{n-1} T_{k-1} = T_{nk}$$

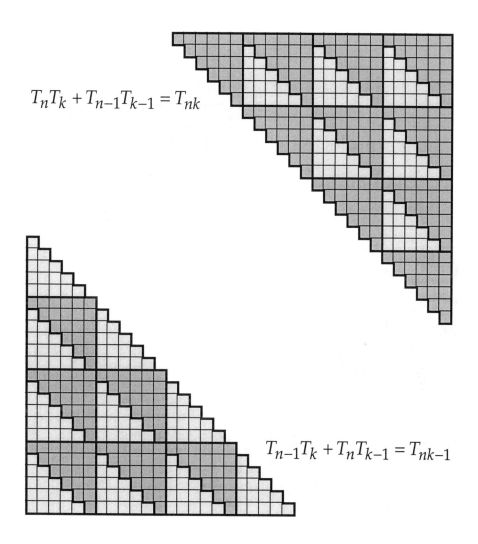

$$T_{n-1} T_k + T_n T_{k-1} = T_{nk-1}$$

—RBN

Identities for Triangular Numbers III

$$T_n = 1 + 2 + \cdots + n \Rightarrow$$

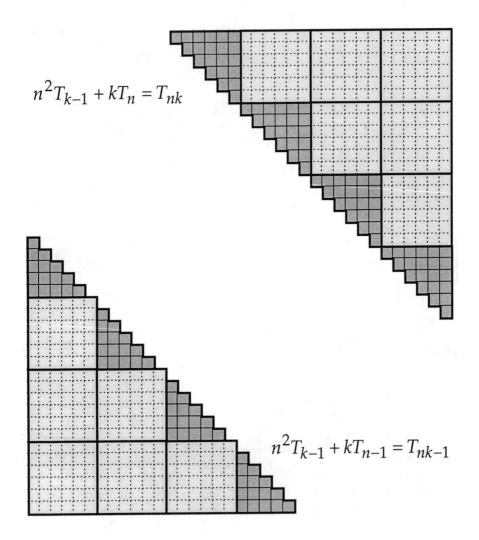

$$n^2 T_{k-1} + k T_n = T_{nk}$$

$$n^2 T_{k-1} + k T_{n-1} = T_{nk-1}$$

—James O. Chilaka

Identities for Pentagonal Numbers

$$P_n = 1 + 4 + 7 + \cdots + (3n-2) \left. \right\}$$
$$T_n = 1 + 2 + 3 + \cdots + n \qquad \Bigg\} \Rightarrow$$

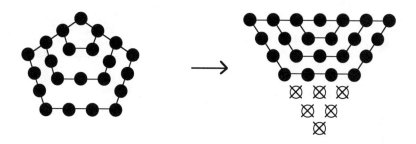

$$P_n = T_{2n-1} - T_{n-1}$$

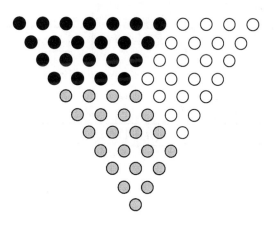

$$P_n = \frac{1}{3} T_{3n-1}$$

Sums of Octagonal Numbers

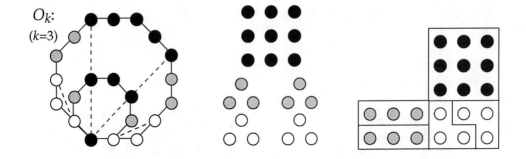

$$T_k = 1 + 2 + \cdots + k \Rightarrow O_k = k^2 + 4T_{k-1}$$

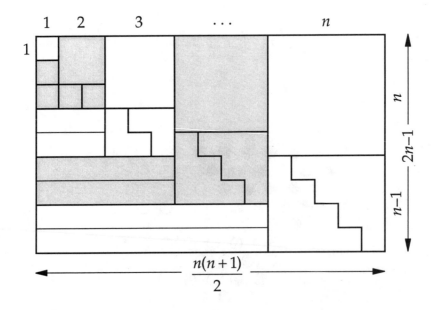

$$\sum_{k=1}^{n} O_k = 1 + 8 + 21 + 40 + \cdots + (n^2 + 4T_{n-1}) = \frac{n(n+1)(2n-1)}{2}$$

—James O. Chilaka

Sums of Products of Consecutive Integers I

$$\sum_{k=1}^{n} k(k+1) = \frac{n(n+1)(n+2)}{3}$$

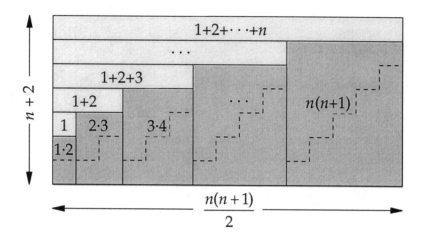

$$T_k = 1 + 2 + \cdots + k \quad \Rightarrow$$

$$1 \cdot 2 + 2 \cdot 3 + \cdots + n(n+1) + \left(T_1 + T_2 + \cdots + T_n\right) = \frac{n(n+1)(n+2)}{2},$$

$$T_1 + T_2 + \cdots + T_n = \frac{1}{2}\left(1 \cdot 2 + 2 \cdot 3 + \cdots + n(n+1)\right),$$

$$\therefore \frac{3}{2}\left(1 \cdot 2 + 2 \cdot 3 + \cdots + n(n+1)\right) = \frac{n(n+1)(n+2)}{2}.$$

—James O. Chilaka

Sums of Products of Consecutive Integers II

$$\sum_{k=1}^{n} k(k+1)(k+2) = \frac{n(n+1)(n+2)(n+3)}{4}$$

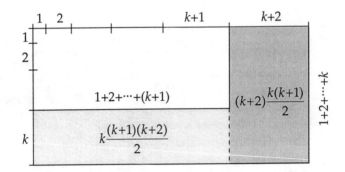

$$k\frac{(k+1)(k+2)}{2} + (k+2)\frac{k(k+1)}{2} = k(k+1)(k+2)$$

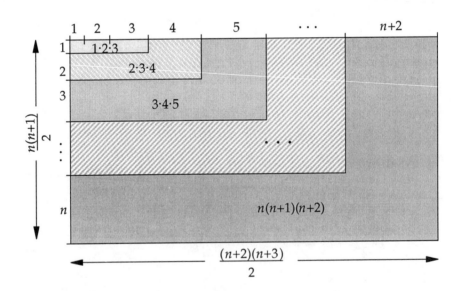

$$1 \cdot 2 \cdot 3 + 2 \cdot 3 \cdot 4 + \cdots + n(n+1)(n+2)$$
$$= \frac{n(n+1)}{2} \times \frac{(n+2)(n+3)}{2} = \frac{n(n+1)(n+2)(n+3)}{4}$$

—James O. Chilaka

Fibonacci Identities

$$F_1 = F_2 = 1, \quad F_n = F_{n-1} + F_{n-2} \Rightarrow$$

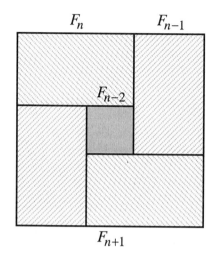

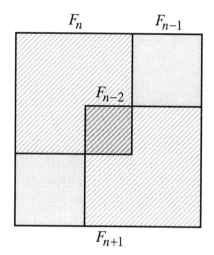

$$F_{n+1}^2 = 4F_nF_{n-1} + F_{n-2}^2 \qquad\qquad F_{n+1}^2 = 2F_n^2 + 2F_{n-1}^2 - F_{n-2}^2$$

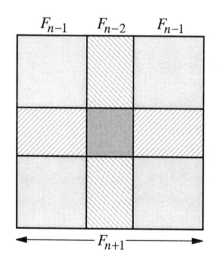

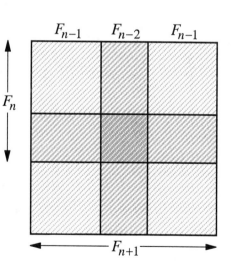

$$F_{n+1}^2 = 4F_{n-1}^2 + 4F_{n-1}F_{n-2} + F_{n-2}^2 \qquad F_{n+1}^2 = 4F_n^2 - 4F_{n-1}F_{n-2} - 3F_{n-2}^2$$

—Alfred Brousseau

Sums of Powers of Three

$$\sum_{k=1}^{n-1} 3^k = \frac{3^n - 1}{2}$$

3^0 3^1 3^2 3^3 3^4 $\cdots$

$$3^n - 1 = 2\sum_{k=1}^{n-1} 3^k$$

—David B. Sher

Infinite Series, Linear Algebra, & Other Topics

A Geometric Series

$$\frac{1}{4} + \left(\frac{1}{4}\right)^2 + \left(\frac{1}{4}\right)^3 + \cdots = \frac{1}{3}$$

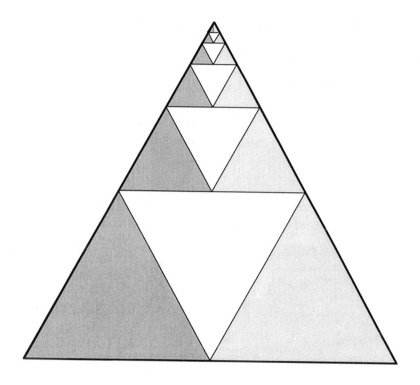

—Rick Mabry

An Alternating Series

$$\frac{1}{2} - \frac{1}{4} + \frac{1}{8} - \frac{1}{16} + \frac{1}{32} - \frac{1}{64} + \cdots = \frac{1}{3}$$

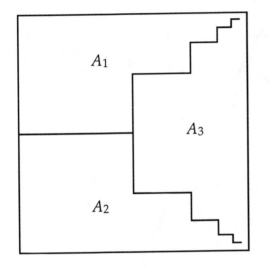 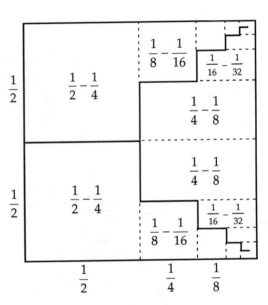

$$A_1 = \frac{1}{2} - \frac{1}{4} + \frac{1}{8} - \frac{1}{16} + \frac{1}{32} - \frac{1}{64} + \cdots,$$

$$A_1 = A_2 = A_3,$$

$$A_1 + A_2 + A_3 = 1,$$

$$\therefore A_1 = \frac{1}{3}.$$

—James O. Chilaka

A Generalized Geometric Series

Let $\{k_1, k_2, k_3, \cdots\}$ be a sequence of integers, each of which is at least 2. Then

$$\frac{k_1-1}{k_1} + \frac{k_2-1}{k_2 k_1} + \frac{k_3-1}{k_3 k_2 k_1} + \cdots = 1.$$

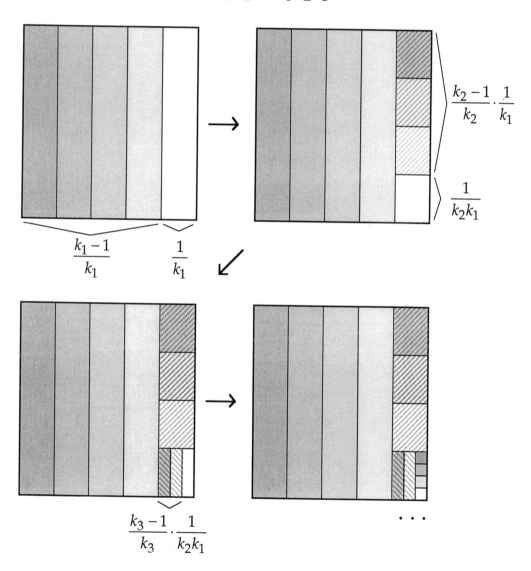

—John Mason

Divergence of a Series

$$n > 1 \Rightarrow \sum_{k=1}^{n} \frac{1}{\sqrt{k}} > \sqrt{n}$$

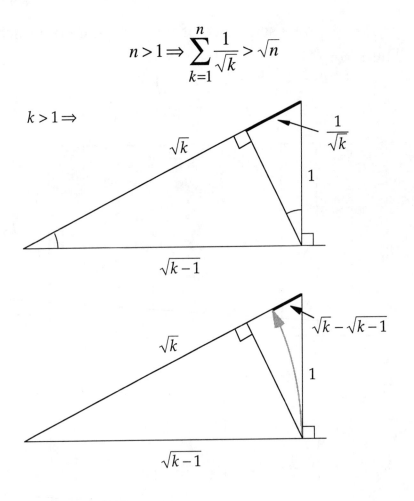

$$\frac{1}{\sqrt{k}} > \sqrt{k} - \sqrt{k-1}$$

$$\frac{1}{\sqrt{2}} + \frac{1}{\sqrt{3}} + \cdots + \frac{1}{\sqrt{n}} > \left(\sqrt{2} - 1\right) + \left(\sqrt{3} - \sqrt{2}\right) + \cdots + \left(\sqrt{n} - \sqrt{n-1}\right)$$

$$\therefore 1 + \frac{1}{\sqrt{2}} + \frac{1}{\sqrt{3}} + \cdots + \frac{1}{\sqrt{n}} > \sqrt{n}$$

—Sidney H. Kung

Galileo's Ratios

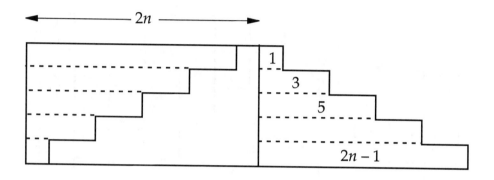

$$\frac{1}{3} = \frac{1+3}{5+7} = \frac{1+3+5}{7+9+11} = \cdots = \frac{1+3+5+\cdots+(2n-1)}{(2n+1)+(2n+3)+\cdots+(2n+2n-1)}$$

—Alfinio Flores

Sums of Harmonic Sums

$$H_k = 1 + \frac{1}{2} + \frac{1}{3} + \cdots + \frac{1}{k} \Rightarrow \sum_{k=1}^{n-1} H_k = nH_n - n$$

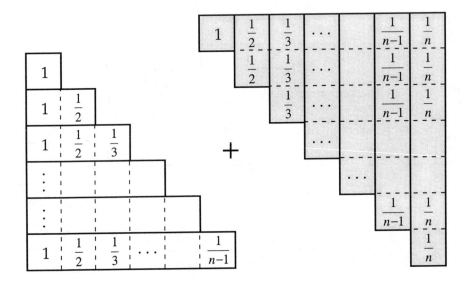

$$\sum_{k=1}^{n-1} H_k + n = nH_n$$

$(\mathbf{A}\mathbf{B})^T = \mathbf{B}^T\mathbf{A}^T$, Where $\mathbf{A}$ and $\mathbf{B}$ are Matrices

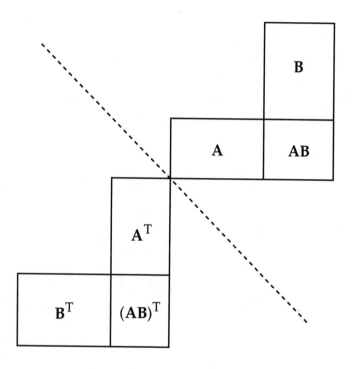

—James G. Simmonds

The Distributive Property of the Triple Scalar Product

$$\vec{A}\cdot\left(\vec{C}\times\vec{D}\right)+\vec{B}\cdot\left(\vec{C}\times\vec{D}\right)=\left(\vec{A}+\vec{B}\right)\cdot\left(\vec{C}\times\vec{D}\right)$$

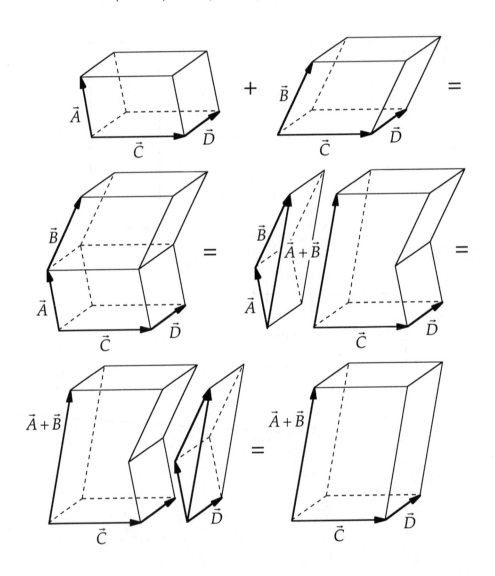

—Constance C. Edwards
and Prashant S. Sansgiry

Cramer's Rule

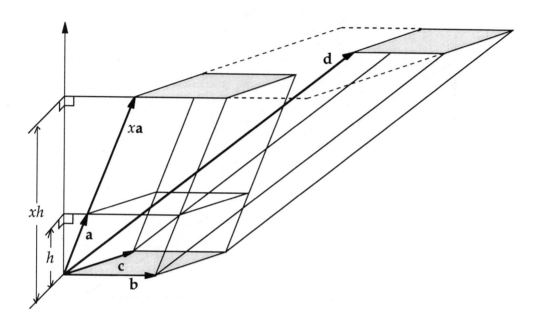

$$xa + yb + zc = d \Rightarrow \det(d,b,c) = \det(xa,b,c) = x\det(a,b,c)$$

$$\therefore x = \frac{\det(d,b,c)}{\det(a,b,c)}$$

—*The Mathematics Initiative,*
Education Development Center

Parametric Representation of Primitive Pythagorean Triples

$$\frac{a}{2}, b, c \in \mathbf{Z}^+, \quad (a,b) = 1$$

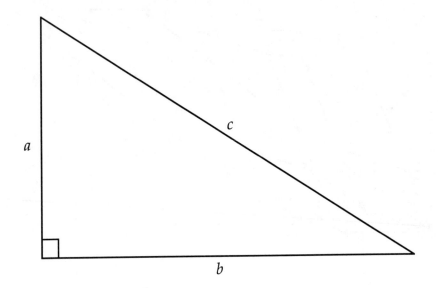

$$\frac{c+b}{a} = \frac{n}{m}, \ (n,m) = 1 \Rightarrow \frac{c-b}{a} = \frac{m}{n},$$

$$\Rightarrow \frac{c}{a} = \frac{n^2 + m^2}{2mn}, \ \frac{b}{a} = \frac{n^2 - m^2}{2mn},$$

$$\Rightarrow n \not\equiv m \,(\mathrm{mod}\,2).$$

$$\therefore (a,b,c) = (2mn, n^2 - m^2, n^2 + m^2).$$

—Raymond A. Beauregard
and E. R. Suryanarayan

On Perfect Numbers

$$p = 2^{n+1} - 1 \text{ prime} \quad \Rightarrow \quad N = 2^n p \text{ perfect}$$

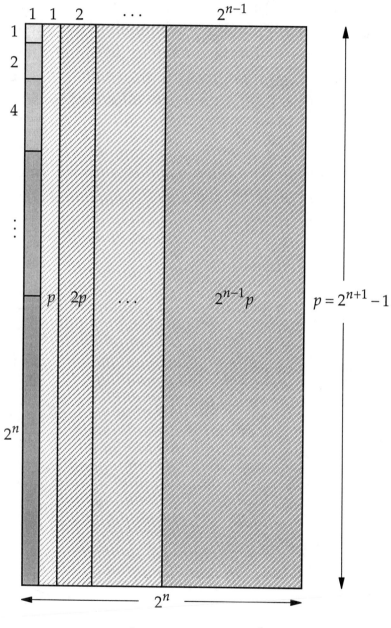

$$1 + 2 + \cdots + 2^n + p + 2p + \cdots + 2^{n-1}p = 2^n p = N$$

—Don Goldberg

Self-Complementary Graphs

A graph is *simple* if it contains no loops or multiple edges. A simple graph $G = (V,E)$ is *self-complementary* if G is isomorphic to its *complement* $\overline{G} = (V,\overline{E})$, where $\overline{E} = \{\{v,w\}: v,w \in V,\ v \neq w,\ \text{and}\ \{v,w\} \notin E\}$. It is a standard exercise to show that if G is a self-complementary simple graph with n vertices, then $n \equiv 0$ (mod 4) or $n \equiv 1$ (mod 4). A converse also holds, as we now show.

THEOREM: *If n is a positive integer and either $n \equiv 0$ (mod 4) or $n \equiv 1$ (mod 4), then there exists a self-complementary simple graph G_n with n vertices.*

PROOF:

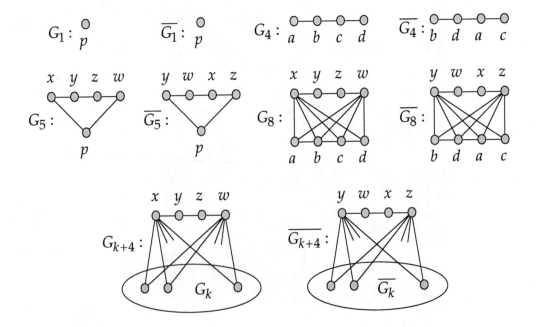

—Stephan C. Carlson

Tiling with Trominoes

A *tromino* is a plane figure composed of three squares: ⌐⌐

THEOREM: *If n is a power of two, then an n × n checkerboard with any one square removed can be tiled using trominoes.*

PROOF (by induction):

I.

II.

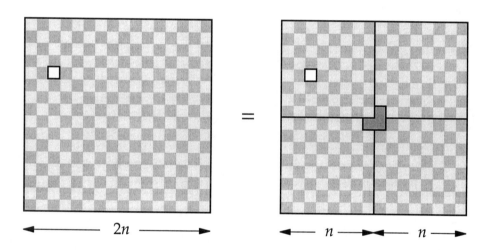

—Solomon W. Golomb

NOTE: Except when $n = 5$, an $n \times n$ checkerboard with any one square removed can be tiled with trominoes if and only if $n \not\equiv 0 \pmod{3}$. See I-Ping Chu and Richard Johnsonbaugh, "Tiling deficient boards with trominoes," *Mathematics Magazine* 59 (1986) 34-40.

Sources

page	source

Geometry & Algebra

3   http://tug.org/applications/PSTricks/Tilings

4   *Mathematics Magazine*, vol. 71, no. 3 (June 1998), p. 170.

5   Howard Eves, *Great Moments in Mathematics (Before 1650)*, The Mathematical Association of America, Washington, 1980, pp. 29-31.

6   Reprinted with permission from Elisha Scott Loomis, *The Pythagorean Proposition*, p. 112, copyright 1968 by the National Council of Teachers of Mathematics. All rights reserved.

7   *College Mathematics Journal*, vol. 27, no. 5 (Nov. 1996), p. 409.

8   *Mathematics Magazine*, vol. 72, no. 5 (Dec. 1999), p. 407.

9   *Mathematics Magazine*, vol. 71, no. 1 (Feb. 1998), p. 64.

10   *Mathematics Magazine*, vol. 70, no. 5 (Dec. 1997), p. 380; vol. 71, no. 3 (June 1998), p. 224.

11   *Mathematics Magazine*, vol. 71, no. 4 (Oct. 1998), p. 314.

12   http://www.cms.math.ca/CMS/Competitions/OMC/examt/english69.html

13   I. *Mathematics Magazine*, vol. 71, no. 3 (June 1998), p. 196.
II. Ross Honsberger, *Mathematical Morsels*, The Mathematical Association of America, Washington, 1978, pp. 27-28.

14   *Mathematics Magazine*, vol. 72, no. 4 (Oct. 1999), p. 317.

15   Written communication.

16   *Mathematics Magazine*, vol. 72, no. 2 (April 1999), p. 142.

17   Reprinted with permission from *The Mathematics Teacher* [vol. 85, no. 2 (Feb. 1992), front cover; vol. 86, no 3 (March 1993), p. 192], copyright 1992, 1993 by the National Council of Teachers of Mathematics. All rights reserved.

18   *American Mathematical Monthly*, vol. 93, no. 7 (Aug.-Sept. 1986), p. 572.

19   *College Mathematics Journal*, vol. 28, no. 3 (May 1997), p. 171.

20   *Mathematics and Computer Education*, vol. 33, no. 3 (Fall 1999), p. 282.

22   Ross Honsberger, *Mathematical Morsels*, The Mathematical Association of America, Washington, 1978, pp. 204-205.

24   *American Mathematical Monthly*, vol. 86, no. 9 (Nov. 1979), pp. 752, 755.

Index of Names

Technical Note

The manuscript for this book was edited and printed using Microsoft® Word 6.0.1 on an Apple® Macintosh™ Power PC computer. The graphics were produced in ClarisDraw™ 1.0v4. The text is set in the Palatino font, with special characters in the Symbol font. Displayed equations were produced with Microsoft® Equation Editor v2.01. The manuscript was printed on a Hewlett Packard® LaserJet 4MV.